Human Factors for Pilots and Aviation Specialists

This book investigates human behavior within the aviation sector. Authored by an experienced aviation psychologist, it reveals the complexities of human performance in flight operations. From understanding the question of how human factors shape aviation safety to exploring decision-making, spatial awareness, stress management, and well-being, this book details what aviation professionals need to know about the industry of flight and psychology.

This title integrates current research, industry practice, and regulatory frameworks to show how human factors and psychological principles influence aviation safety and performance. It offers experience-based insights into the psychological dimensions of flight operations, fostering a deeper understanding of how human performance shapes safety and efficiency in modern aviation. Drawing on more than forty recent real-world case studies and operational examples, the book bridges theory and practice through practical takeaways and reflective questions that support confident, applied decision-making across aviation roles.

Human Factors for Pilots and Aviation Specialists: Flight Safety Through Modern Approaches to Human Behavior will appeal to experienced aviation professionals, human-factors specialists, students of aerospace and aeronautical engineering and researchers studying the human aspects of flight.

Human Factors for Pilots and Aviation Specialists

Flight Safety Through Modern Approaches to Human Behavior

Aleksandra Kapela

CRC Press
Taylor & Francis Group
Boca Raton New York London

CRC Press is an imprint of the
Taylor & Francis Group, an **informa** business

Designed cover image: Shutterstock

First edition published 2026
by CRC Press
2385 NW Executive Center Drive, Suite 320, Boca Raton FL 33431

and by CRC Press
4 Park Square, Milton Park, Abingdon, Oxon, OX14 4RN

Library of Congress Cataloging-in-Publication Data
Names: Kapela, Aleksandra author
Title: Human factors for pilots and aviation specialists : flight safety through modern approaches to human behavior / Aleksandra Kapela.
Description: First edition. | Boca Raton FL : CRC Press, 2026. | Includes bibliographical references.
Identifiers: LCCN 2025051912 | ISBN 9781032848150 hardback | ISBN 9781032848723 paperback | ISBN 9781003515418 ebook
Subjects: LCSH: Aviation psychology | Air pilots--Psychology | Aeronautics--Safety measures
Classification: LCC TL555 .K37 2026
LC record available at https://lccn.loc.gov/2025051912

ISBN: 978-1-032-84815-0 (hbk)
ISBN: 978-1-032-84872-3 (pbk)
ISBN: 978-1-003-51541-8 (ebk)

DOI: 10.1201/9781003515418

Typeset in Times
by KnowledgeWorks Global Ltd.

For my daughter, Rose – may you always know that there are no limits to what you can achieve.

Contents

About the Author

Aleksandra Kapela is Founder of MyWingman.eu an aviation psychology initiative focused on supporting pilot training, assessment, and wellbeing. She is an aviation psychologist and human-factors specialist with over a decade of experience supporting pilots and aviation organizations worldwide. She holds master's degrees in Occupational Psychology and Human Resource Management and is also a ground school instructor involved in pilot training. Certified by the European Association for Aviation Psychology (EAAP), she specializes in competency-based training, KSA100 assessment, and crew performance. Ms. Kapela regularly speaks at international aviation conferences and works with airlines, training organizations, and universities to integrate psychological science into aviation training and safety culture. Her work focuses on bridging operational aviation with modern insights from psychology, neuroscience, and human performance.

Foreword

Aviation is a contradiction.

In 2023, commercial aviation generated approximately US $4.1 trillion in global economic impact – around 3.9% of world GDP. If aviation were a country, its contribution would rank it among the four largest economies on Earth.

Aviation is also extraordinarily safe. In the same year, 4.4 billion passengers flew on nearly 29,000 aircraft, completing around thirty-eight million flights, without a single fatality in a commercial jet aircraft. Yet this statement, while factually correct, is incomplete.

Aviation is not safe because accidents are rare. It is safe because, every day, thousands of professionals detect, absorb, correct, and recover from small failures before they escalate. Safety is not an outcome. It is a habit. A culture. Something people do when no one is watching – especially when things go wrong.

Why do intelligent people fear flying? Given the transparency of aviation safety data, a troubling question remains: why do nearly 20% of the population fear flying, including some of the most intelligent, analytical, and experienced professionals? Why do around 8% choose not to fly at all, opting instead for ground transport that is orders of magnitude more dangerous?

I encounter this contradiction frequently. Leaders who confidently calculate risk, champion artificial intelligence, and speak fluently about resilience will quietly admit they fear boarding an aircraft. The problem is not ignorance of statistics. It is a failure to understand how safety works.

Resilience is not a single skill. Many leaders speak about resilience as if it were a standalone trait: knowledge, training, experience, teamwork, leadership, decision-making, crisis management, and risk. Resilience is none of these in isolation. True resilience emerges only when they are interlinked through communication, trust, and shared understanding of our capabilities and limitations.

> A founder with a high IQ but poor emotional intelligence is not resilient.
> A technically brilliant expert who cannot regulate stress or work in a team is not resilient.

Resilience is systemic. Human Factors for Pilots and Aviation Specialists (HFPAP) bridges the gaps between technology, complexity, risks, culture, safety, and resilience. It explains how people remain calm in crises while others lose cognitive control. It is both a technical and human guide to maintaining physical and mental performance when the unthinkable occurs.

Professionalism, as this book makes clear, is not blind obedience. It is understanding systems to their foundations, why each step exists and knowing when breaking the rules is required. In an increasingly complex environments where automation does most of the work but takes none of the responsibility, it is up to the pilots to save the day, a team alone, often with no way to stop or call a friend.

Aviation remains safe because professionals expect errors, respect them and then design for them. Stress cannot be eliminated, but it can be regulated. Resilience is the ability to prepare during calm periods so that teams can think, communicate, and decide effectively when things unravel. It is not merely about surviving adversity; resilient systems thrive during stability and perform under strain.

Resilience is not luck. It is preparation, teaming, and execution. Across its chapters, this book explores how attention, workload, situational awareness, and crisis decision-making evolves in real operations. It explains why situational awareness can appear stable until it collapses suddenly, why underload can be as dangerous as overload, and why experience can be a curse.

Aviation regulations are often described as "written in blood." But safety is sustained not by rules alone – it is sustained by people. The professionals we entrust with our lives do not simply follow SOPs in calm conditions; they possess the skills to create novel solutions when faced with unprecedented failures that no checklist can resolve.

To succeed in aviation – and in any high-risk domain – it is not enough to understand "what" to do. One must understand the "how" and the "why." This is what Human Factors really means.

I define Human Factors as studies of four interconnected dimensions:

- How humans are built
- What humans can do
- How humans respond to inputs
- How humans interact with machines

To grasp these, we must understand the neuroscience of the brain and mind – not just neurons and synapses, but sensory inputs, motor outputs, emotion, stress chemistry, and cognition. This knowledge allows us to connect every element of resilience into a coherent whole.

Human limits are not character flaws. They are biological facts. Understanding them gives us the confidence to own our limitations, explain our decisions, and shape outcomes deliberately. Human Factors teaches us how to harness both our strengths and our vulnerabilities, to mitigate threats, survive, and lead, through crises.

Leadership when the tide goes out. Fair-weather leaders can succeed during calm periods. But when pressure rises and margins disappear, the truth is revealed. As Warren Buffett famously observed, "Only when the tide goes out do you discover who's been swimming naked."

Sustained resilience belongs to those who understand – and respect – human factors.

Why this book matters. For my entire life, I have advocated continuous learning, adaptation, and improvement. I am a strong proponent of Human Factors and resilience – not as theory, but as a necessity for life.

That is why I am delighted that Aleksandra Kapela asked me to write this foreword. She is among the very few who understand Human Factors – and rarer still, written about it with clarity, depth, and operational relevance.

By exploring all four dimensions of Human Factors, this book explains how we think, act, communicate, and build resilience. In just over 122 years since the first powered flight, aviation has evolved from the most dangerous form of transport to the safest. The lessons in this book explain the WHYs, the HOWs, and the WHATs.

They are lessons not just for aviation, but for personal, corporate, and national resilience.

Human Factors for Pilots and Aviation Specialists is not a manual of procedures. It is a human performance book for people who operate in complex, high-risk systems – where failure is not an option, uncertainty is constant, and comfort is rare.

I commend this book to you.

Captain Richard Champion de Crespigny
Pilot-in-Command, Qantas Flight QF32
4 November 2010

Preface

This book offers a fresh, modern view of human performance in aviation. This may be the first book in aviation to bring together **academic insight, neurocognitive foundations, and a practical operational perspective –** making human factors both understandable and useful in everyday practice. Drawing on 40 real-world contemporary case studies from 2009 onwards, it explores how people think, decide, and perform under pressure – focusing primarily on pilots and instructors while recognizing the vital interdependence of all aviation roles, from cabin crew and engineers to air traffic controllers. These case studies include both lessons learned from setbacks and inspiring examples of resilience and good practice that have advanced modern aviation safety.

This book grew out of years spent listening to professionals across the flight deck, the cabin, and offices, all sharing the same curiosity about what truly shapes human performance. Written in clear, accessible language, it links scientific research with operational experience, showing how stress, fatigue, attention, leadership – but also mental health and well-being – influence safety and teamwork across the system.

It also looks ahead to emerging technologies, training innovations, and future human–machine collaboration, offering a holistic, forward-looking understanding of human factors for the next generation of aviation professionals.

By blending evidence, empathy, and forward-looking insight, this book invites everyone passionate about aviation to rethink safety, performance, and the human mind in flight. My hope is that it helps current and future professionals – pilots above all, yet every role beside them – see human factors not as a checklist, but as the living heartbeat of every safe and meaningful flight.

Acknowledgments

Writing this book has been like a long flight – full of planning, turbulence, quiet reflection, and unexpected tailwinds. I owe many people for helping me reach this destination.

My deepest gratitude goes to **Capt. Richard de Crespigny**, who generously contributed the foreword to this book and shared thoughtful feedback during its development, and to **Capt. Tammie Jo Shults**, whose openness and kindness in our exchanges embodied the kind of leadership and composure that inspired much of this work.

To the many aviation professionals I had the privilege to speak with while researching the case studies - pilots, cabin crew, air traffic controllers, engineers, and investigators - thank you for sharing your time, insight, and honesty. Your experiences brought depth, accuracy, and humanity to every story in this book. I am also grateful to **Petter Hörnfeldt (Mentour Pilot)** for generously sharing his insights on the case studies discussed in this book.

To my professional network and colleagues across aviation - psychologists, instructors, and human-factors and safety specialists - thank you for every conversation, challenge, and spark of motivation that turned years of notes and shared experiences into these chapters. I am especially grateful to **Bartho Blom**, whose constant support, candid challenges, and deep industry insight helped sharpen many of the ideas explored in this book. His willingness to question assumptions and engage in honest discussion made this work stronger.

To the team at **Taylor & Francis**, thank you for believing in me and offering me the opportunity to turn my passion into a book. I'm especially grateful to those who guided and supported me throughout the process - your encouragement, patience, and trust made this journey both possible and deeply rewarding.

To my dear family— my sister, parents, husband — and my friends: thank you for your patience, support, encouragement, and for blowing wind into my wings whenever they felt heavy. Your faith and help kept me going when the finish line seemed far away.

And finally, to every aviation enthusiast and reader who continues to learn, question, and grow in this field – thank you for caring about the human side of flight. You are the reason human factors keep evolving.

Acknowledgments

1 The Human in the System
From Blame to Complexity in Aviation Safety

CASE STUDY: THE TENERIFE ACCIDENT (1977) – MORE THAN JUST A HUMAN ERROR

It began with something that wasn't even supposed to happen – a diversion.

On March 27, 1977, a terrorist bombing at Gran Canaria Airport forced several flights to divert. Among them were two Boeing 747s – one operated by KLM Royal Dutch Airlines (KLM) and the other by Pan Am (ICAO, 1978, 1984).

Los Rodeos Airport on Tenerife was a small regional field, never meant for that kind of traffic. Taxiways became parking areas, and the ramp filled up with widebodies waiting for clearance. Controllers were overloaded, crews sat for hours with engines shut down, and fatigue and impatience began to creep in.

Then the fog rolled in. Visibility dropped to almost nothing. From the tower, the controllers could no longer see the runway; from the cockpits, the pilots could barely see past the wingtips (ICAO, 1978, 1984).

In the KLM cockpit sat Captain Jacob Veldhuyzen van Zanten – one of the airline's most senior training captains, known for precision and professionalism. But it was late in the duty day, the crew were watching their flight-time limits, and the captain was anxious to get home. The Pan Am crew, meanwhile, were still taxiing down the same runway, trying to find the correct exit. Without ground radar, nobody knew exactly where anyone was.

The moment that sealed the tragedy was almost mundane – a radio call. The KLM captain believed he had been cleared for takeoff. His first officer repeated words that sounded close enough to a clearance to feel like confirmation, though the tower had not given one. The flight engineer questioned it – was Pan Am still on the runway? – but the captain pressed the throttles forward (ICAO, 1978; Weick, 1990).

Pan Am was still there. From their perspective, the situation was shaped by a series of ambiguous and changing instructions. In dense fog and without ground radar, they were instructed to vacate the runway via an early exit whose identification and geometry were challenging for a Boeing 747. Communication with air traffic control required repeated clarification, including confusion around taxiway numbering, while controller workload was unusually high due to traffic congestion and limited staffing. The Pan Am crew reported that they

DOI: 10.1201/9781003515418-1

were still taxiing on the runway, but these transmissions were either blocked or not received by the KLM cockpit. With no indication that a take-off clearance had been issued behind them, and with prevailing communication practices at the time allowing for informal acknowledgements rather than strict readbacks, the crew reasonably assumed that air traffic control retained full situational awareness of their position (ICAO, 1978).

Seconds later, the two 747s collided in the fog, erupting in fire and debris. Five hundred and eighty-three people lost their lives. It remains the deadliest accident in aviation history (ICAO, 1978, 1984, 2023).

At first, the conclusion seemed simple: a pilot error. The captain took off without clearance. Case closed. But Tenerife became the turning point – the moment aviation realized that safety is never about one person, one error, or one bad decision. It was about fatigue, time pressure, ambiguous communication, the absence of radar, heavy workload, authority gradients, and a cockpit culture that made it hard to speak up (ICAO, 1984; Weick, 1990).

It marked the beginning of a new way of thinking – the start of human factors as we know it today.

INTRODUCTION: FROM WHERE WE STARTED TO WHERE WE'RE HEADING

Today, few see Tenerife as a story of one bad decision. We see it for what it was: a system that failed; a mix of stress, hierarchy, ambiguous language, fatigue, and a cockpit culture that used to value silence over questioning.

Listening to the recordings, you don't hear recklessness, but humans juggling too much at once. They were wrong, but not careless. It was human behavior under extraordinary pressure. That's what made Tenerife a turning point: it revealed how fragile communication can be, how authority distorts clarity, and how even the most seasoned crews can slip when the system around them doesn't catch the error.

Tenerife marked a shift: from blaming people to understanding them, from punishing mistakes to learning from them. That change gave rise to Crew Resource Management (CRM), planted the seeds of modern human factors, and began a conversation we're still having – because we must. For everyone in aviation: pilots, cabin crew, controllers, instructors, engineers, the story still echoes. It reminds us that neither experience nor technology makes us immune. Not because people are careless but because they are human.

This book is about those humans: how they think, act, adapt (and sometimes fail), in one of the most unforgiving workplaces in the world: the sky.

After Tenerife came a deeper realization: "human error" explains little. It only gives us someone to blame. Slowly, we began to see that people were not just fallible but adaptable and resilient – often holding things together with limited tools and almost no time (Dekker, 2014). Then came another insight: to study safety, we must also study success. Erik Hollnagel called it "Safety-II" – the study of what goes right (Hollnagel, 2014).

Those quiet recoveries rarely make headlines but matter deeply: crews stabilizing workload, engineers catching faults, controllers managing near-overload. Humans are not just sources of risk; they are often the reason safety holds.

To understand performance today, we must look ahead as much as back – drawing on behavioral science, cognitive psychology, and operational experience (Wickens et al., 2022). Yet knowledge brings no certainty. No simulator captures the chaos of a real emergency: the noise, the narrowing focus, the weight of the unknown (Martin et al., 2016; Swauger, 2023).

That is why aviation can't afford a culture of blame. Growth comes from reflection, not shame. From asking: *What would I have done? What don't I know yet?* Instead of *Whose fault is it?*

This chapter begins there – at the moment aviation stopped blaming people and started understanding systems.

So let's dive in.

FROM BLAME TO SYSTEMS THINKING

For decades, aviation searched for a name – the individual behind the mistake.

If you could find the person, you could fix the problem. It felt simple, even satisfying.

But over time, we learned that people don't fail in isolation. They fail *within* systems, and often because those systems quietly set them up to.

That shift, from *who* to *why*, is the pivot on which modern safety thinking turns (Dekker, 2014; Reason, 1990). The **system thinking** recognizes that the environment, design, workload, culture, and fatigue all shape what professionals see, decide, and do under pressure. When a flight goes wrong, the useful question isn't *Who was at fault?* but *How did this make sense at the time?*

This perspective rebuilt aviation safety. It underpins CRM, feeds Threat and Error Management (TEM) and lives within Safety Management System (SMS) – where reporting is treated not as confession but as intelligence (ICAO, 2018; Kanki et al., 2019).

It doesn't excuse error; it redirects accountability toward the conditions that made it possible.

The Person-Centered View – Errors as Personal Failures and Limitations of Blame

Early aviation borrowed the military model: steep hierarchy, strict discipline, personal responsibility. When something broke, someone was to blame.

That mindset ignored how human performance actually works. Under high workload, the brain narrows focus, filters cues, and relies on assumptions. Memory fades, attention tunnels, and the chemistry of stress erodes reasoning (Arnsten et al., 2015; Miller, 1956; Wickens et al., 2022).

Most errors don't come from negligence; they come from biology. Fatigue slows reaction time. Stress impairs inhibition. Automation bias and confirmation bias make the improbable invisible. The **fundamental attribution error** – our bias to explain others' behavior by who they are rather than the situation they're in – fed years of person-blame investigations (Ross, 1977). The result was silence: fewer reports, hidden near-misses, and uncorrected systemic flaws.

A Picture That Changed the Conversation

James Reason's Swiss Cheese Model gave safety a language. It showed that accidents are rarely single points of failure – they're alignments across layers of defense: design, training, supervision, and operation (Reason, 1990, 1997). Each layer has holes; most days those holes don't line up. On the bad day, they do.

Reason also distinguished between **active failures** – the actions of those on the front line – and **latent conditions**, the hidden weaknesses in systems such as vague Standard Operating Procedures (SOPs), fatigue-inducing schedules, or unclear communication culture. Latent conditions don't crash aircraft by themselves, but they set the stage for failure. The message was clear: error is inevitable, but its *consequences* are not. (Reason, 1990)

In this view, safety's aim isn't to eliminate error but to detect, trap, and recover from it. Figure 1.1 applies this logic to Tenerife: diversion pressure, fatigue, ambiguous phraseology, dense fog, missing defenses, and a hierarchy too steep to challenge. The event wasn't one bad decision; it was multiple small weaknesses aligning at once.

This marked a permanent shift – from "Who failed?" to "What made this failure possible?," redirecting focus from individual blame to the conditions that made the outcome likely. It remains the foundation of modern investigation and SMS, where reports are treated as foresight, not confession (Reason, 1990).

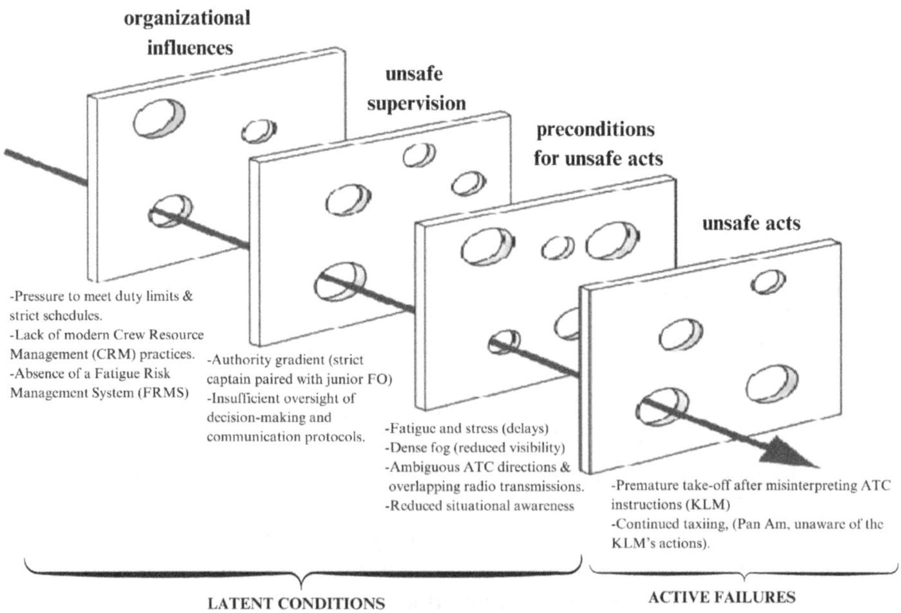

FIGURE 1.1 Swiss Cheese Model Applied to the Tenerife Disaster.

Just Culture and the Learning Mindset

As systems thinking matured, another truth surfaced: people will not speak up unless it is safe to do so. A **just culture** encourages professionals to report incidents, near-misses, and even their own doubts without fear of unfair punishment – while still holding willful violations to account (Dekker, 2012). It is not the removal of accountability. It is the calibration of it.

When crews trust the system, they share. That data are the lifeblood of predictive safety. It is how organizations spot weak signals before they turn into headlines. Just culture also protects health. When people can raise fatigue, stress, or workload concerns early, they are more likely to stay well and perform well (ICAO, 2020, 2018). This concept is explored in more depth in the next chapter: *Safety Culture: Beyond Compliance – A Safety Mindset.*

Systems Thinking in Practice

Systems thinking shifts focus from the individual at the sharp end to the environment shaping their decisions. Its true test lies not in theory but on the line, when small drifts are caught before they grow.

> Fog pressed low over Austin on February 4, 2023. A **Federal Express (FedEx) Flight 1432** was on approach to runway 18L while a **Southwest Airlines Flight 708** lined up for departure on the same strip. The controller's plan seemed safe (separation should have been enough) but lineup took longer than expected. In the thick morning fog, no one could see how close the two paths were about to cross. Austin had no surface-movement alerting system, and the tower's view was almost gone.
>
> The FedEx crew broke through the cloud base and saw the Southwest still on the runway. They called the tower, executed a missed approach. At that moment the FedEx aircraft had just crossed the runway threshold while the Southwest aircraft was accelerating down the runway. The aircraft passed within roughly 150 to 170 feet of each other before separation increased. No injuries, no damage, but every safety layer was tested.
>
> The National Transportation Safety Board (NTSB) later noted that what nearly failed were visibility, timing, and missing technological barriers. What saved the day were vigilance, assertive communication, and a rehearsed escape embedded in memory and muscle. The event became a modern reminder that safety is not static; it depends on layers that must be constantly tended, reinforced, and sometimes rebuilt after close calls (NTSB, 2024).

That's systems thinking in action. The question isn't *Who messed up?* but *Why did this seem reasonable at the time?* Which defenses were thin, and which worked? And where should the next investment go – procedures, staffing, or technology? (Hollnagel, 2014; ICAO, 2018).

We now know that aviation safety isn't the absence of incidents. It's the presence of strong, overlapping defenses that help humans recover when conditions degrade –

clear procedures, legible design, scenario-rich training, and operations that manage tempo before it spikes (Hollnagel et al., 2015; ICAO, 2018).

This systems view has quietly reshaped how aviation trains, debriefs, and assesses. The question is no longer simply *Did they run the checklist?* but *What cues were available? How were they building their mental model? Did culture or workload shape the response?*

Competency-Based Training and Assessment (CBTA), CRM, and Fatigue Risk Management Systems (FRMS) share one truth: context shapes behavior. When something goes wrong, we look not only at what the crew did but at what made that action make sense (IATA, 2023).

Designing with this mindset means building systems that support human performance instead of tolerating human error. Pilots are not the last line of defense – they are part of the system itself. Design, culture, and training must help them succeed, not just survive.

THE BIRTH AND EVOLUTION OF CRM

As human factors moved from theory into the cockpit, a practical question emerged: how do we shape the way crews talk, listen, lead, and back each other up when things get tight? If systems thinking helped us understand why accidents happen, CRM gave us a way to change what happens next – in the simulator, on the line, and in the heat of an approach that isn't lining up like it should.

From Tenerife to NASA: The Spark Becomes a Program

Tenerife didn't just expose a communications failure; it forced the industry to look at the social physics of the cockpit – why smart people sometimes stay silent, why a junior hesitates to challenge a senior, and how ambiguity grows as workload rises.

Two years later, NASA Ames brought together researchers, regulators, and airline leaders for the "Resource Management on the Flightdeck" workshop. It marked a psychological pivot. The focus shifted from procedures alone to people in context: leadership, communication, monitoring, and the mechanics of challenge and response. Out of that meeting came the first wave of Cockpit Resource Management – soon broadened to CRM to include cabin crew and others who influence flight safety (Cooper et al., 1980; Helmreich and Merritt, 2001).

United Airlines' early 1980s program paired CRM with Line-Oriented Flight Training (LOFT) so crews could practice communication and decision-making under time pressure. Other carriers followed quickly, embedding the approach globally (Kanki et al., 2019).

The NASA work also gave crews a shared vocabulary for the human side of operations. Terms that once sounded soft – situational awareness (SA), workload management, decision-making under uncertainty, assertiveness, and shared mental models – became operational. None of these live in the margins anymore. They are core competencies: observable, coachable, and measurable.

CRM marked a quiet revolution. It shifted the focus from preventing what goes wrong to building what goes right – confidence under load, clean talk when seconds matter, and small corrections before small drifts grow. The philosophy kept evolving with science – cognition, social psychology, and accident investigation – rather than freezing as a one-era doctrine.

The Generations of CRM – How the Idea Grew Up

CRM did not appear fully formed. It evolved through several generations, each responding to lessons from accidents, new evidence, and operational change. From its early focus on breaking cockpit silence to its present integration within SMS and CBTA, CRM matured into a global standard for teamwork and leadership in aviation. The timeline in Figure 1.2 shows this progression.

Authority, Voice, and the Clean Challenge

The hardest habit to change was not technical but cultural – deference. The solution wasn't to dilute command but to strengthen leadership: approachable authority, explicit invitations to speak, and monitoring that treats challenge as routine (Fabre et al., 2021; Helmreich and Merritt, 2001). A cockpit culture changes only when the least empowered person has a safe, recognized way to intervene.

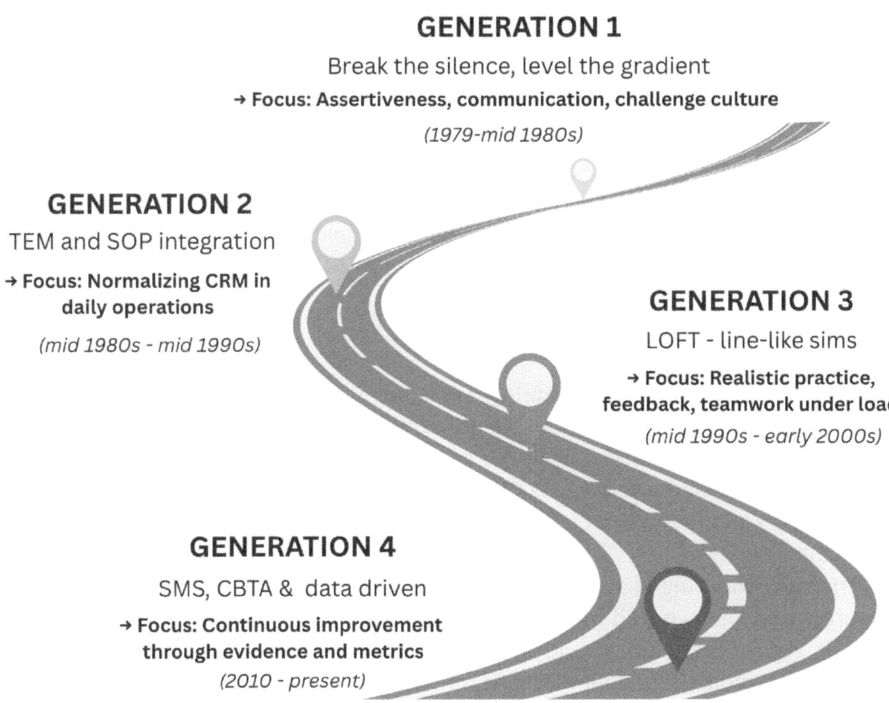

FIGURE 1.2 CRM generations timeline.

CRM made that practical with simple language that lowers the bar for action. When risk rises, one clear statement can stop the chain:

> Captain, I'm not comfortable with this. I recommend we go around.

Sometimes, one honest sentence is all it takes to break the chain.

When the picture starts to blur, crews should use brief, respectful, and specific phrase or pattern – to bring it back into focus. It's anchored to data, not personality, and happens early, not heroically late (Fabre et al., 2021). This turns assertiveness into a disciplined, trainable behavior. It teaches that speaking up is not confrontation but a core flight duty.

Communication under Pressure

Under pressure, communication compresses. People speak less, hesitate more, and fill gaps with assumptions. Shared understanding can evaporate in seconds – exactly when it matters most (Kanki et al., 2019). Early CRM turned this human tendency into trainable countermeasures – structured habits that hold the team together when attention narrows.

- Closed-loop communication ensures clarity: confirm, read back, and expect acknowledgment (UK CAA, 2002)
- Task clarification defines who will do what and by when, preventing overlap and confusion (Helmreich and Merritt, 2001).
- Cross-checking surfaces assumptions so silence isn't mistaken for agreement (Wiener and Nagel, 2010).

These tools are simple by design – and that's why they work. They give crews a common rhythm that holds under stress, distraction, and fatigue. Most importantly, they turn teamwork into something visible, coachable, and assessable in both training and line operations (ICAO, 2018; Kanki et al., 2019).

Lasting Legacy

CRM's strength lies in its adaptability. What began as a reaction to Tenerife became the language of normal operations. Its principles now shape selection, training, and assessment across aviation roles. The core lesson remains the same: communication, leadership, and shared awareness are not soft skills. They are operational safeguards – and they keep evolving with each new generation of crews who practice them.

The most enduring legacy of CRM is cultural. It changed how people see authority, error, and teamwork. It gave aviation a common behavioral language that crosses aircraft types, generations, and national borders. Decades later, it still grows – adapting to automation, data-driven training, and the emerging challenges of modern operations – proof that the human element is not the weakest link, but the one that keeps the system alive.

THE RISE OF COGNITIVE SCIENCE IN AVIATION

CRM gave aviation a behavioral language. Cognitive science explained why it works. Understanding how humans process information, form mental models, and make sense of uncertainty became the next frontier of human factors.

FROM CHECKLISTS TO COGNITION

By the late 1980s, safety experts began asking deeper questions. Why did some crews catch a problem seconds earlier than others? Why did the same procedures sometimes work perfectly and sometimes fail? The answers pointed beyond training or culture to the workings of the mind itself.

Aviation had long focused on what people do. Cognitive science looked at how they think while they do it. The shift revealed a simple truth: procedures may be identical, but perception and attention never are. Fatigue, workload, emotion, and expectation can distort what people see, hear, and recall (Endsley, 2018; Wickens, 2008).

Flying is a cognitive act long before it is a physical one. The brain filters and interprets streams of data into a coherent mental model of what's happening. When that model fits reality, performance looks effortless. When it drifts, small mismatches accumulate until the crew's picture and the aircraft's state no longer align.

Much of aviation psychology since the 1990s has revolved around situational awareness – the ability to perceive, comprehend, and project what's happening in time and space (Endsley, 1995, 2018). SA depends on accurate mental models – internal maps that help predict what should happen next. But models are imperfect. They fill gaps with expectation, masking warning signs.

Accident analysis shows that crews rarely "lose awareness" in an instant. SA erodes gradually – a missed cue here, a misheard call there – until the internal picture no longer matches the external one (Dekker, 2002; Endsley, 2018).

Cognitive science reframed this erosion as a process, not an event. It explained why procedures and teamwork alone aren't enough. People don't just need to follow checklists – they need to understand the mental flow behind them. Under pressure, attention narrows, a phenomenon known as **attentional tunneling** (Wickens, 2008). It aids focus but can make vital cues fade. Stress and fatigue magnify the effect. Adrenaline briefly sharpens focus, then erodes working memory and flexibility (Arnsten et al., 2015). Fatigue slows decision speed and increases rigidity (Van Dongen et al., 2006; Dawson and Reid, 1997).

The insight was simple but profound: the goal isn't perfect attention, but systems, procedures, and teamwork that make lapses visible – and recovery possible.

MANAGING COGNITIVE LOAD

The cognitive load theory became central to understanding why performance fluctuates even in highly trained professionals. Humans can process only a limited number of information channels at once (Miller, 1956). When new data exceed that limit, performance doesn't decline gradually – it collapses.

In a high-workload cockpit, this means critical cues may go unnoticed or simple actions may take longer than expected. Effective design separates essential from non-essential tasks and distributes workload across crew or automation.

CRM behaviors like callouts, monitoring, and cross-checking now made new sense through the lens of cognitive science. They weren't just interpersonal habits; they were externalized thinking strategies that shared mental load between humans.

These insights reframed error recovery as part of normal human performance. The emphasis shifted from eliminating mistakes to building resilience – detecting, trapping, and correcting them before they cascade.

Decision-Making and the Human–Automation Partnership

As automation advanced, researchers turned to naturalistic and recognition-primed decisions (Klein, 1999). **naturalistic decision-making (NDM)** – how experts make fast, high-stakes decisions without consciously comparing options. Experienced operators rely on pattern recognition and mental simulation, a process called the **Recognition-Primed Decision (RPD) model**. A captain doesn't calculate probabilities in the heat of a go-around; they intuitively match cues to mental prototypes formed through experience.

Expertise, however, comes with risk. When mental models become too strong, **confirmation bias** sets in – people see what they expect to see and discount data that contradict it (Reason, 1990). Decision-making under pressure, therefore, balances recognition with verification. Modern training now explicitly builds this flexibility, exposing crews to novel, ambiguous, or counterintuitive scenarios to stretch their pattern library. LOFT and CBTA exercises all draw on this principle (Kanki et al., 2019).

At the same time, cognitive science revealed a new frontier: the **partnership between humans and automation**. The more reliable systems became, the more they invited trust – sometimes excessive trust. **Automation complacency** and **automation bias** emerged as cognitive hazards (Parasuraman and Riley, 1997).

Pilots began relying on systems not just to assist, but to interpret reality for them. When automation behaved unpredictably, humans were often too slow to intervene because their mental model had shifted from "operator" to "supervisor." This imbalance highlighted a key truth: automation and humans form **joint cognitive systems** (Hollnagel and Woods, 2005). When well designed, these systems amplify human strengths and offload routine cognitive burden. When poorly designed, they obscure feedback, slow reaction, and weaken SA.

On August 3, 2016, **Emirates Flight 521** attempted a go-around after a late touchdown at Dubai. The crew pressed the Take-Off/Go-Around (TO/GA) switch expecting an increase in thrust; however, the aircraft had already transitioned into ground mode, in which autothrottle behavior differed from their expectation. As a result, thrust remained low while the aircraft became briefly airborne before settling back onto the runway. The aircraft impacted, slid approximately 800 meters, and caught fire. A firefighter was fatally injured during the response; miraculously, no passengers lost their lives. The investigation identified delayed recognition of the aircraft's energy state, automation mode assumptions, and monitoring breakdowns. The event illustrates how normal, correctly functioning automation can become hazardous when system logic and crew expectations diverge, particularly during time-compressed, high-workload situations (AAIS, 2020).

The solution usually lies in design transparency. Good automation makes its intent visible, communicates its limits, and keeps the human brain in the loop. The aim isn't to replace cognition, but to extend it. In the case of Emirates 521, the automation was not malfunctioning, nor was it behaving unpredictably. It was operating exactly as designed. The problem lay in the opacity of its mode logic at a moment when time, attention, and cognitive capacity were severely constrained. When system states change silently - without salient cues that match human expectations - even well-trained crews may act on incorrect assumptions. This is not a failure of automation, but a reminder that automation design must be evaluated not only for technical correctness, but for cognitive transparency under stress.

The Learning Brain – From Cognition to Competence

Cognitive science didn't just explain why people make errors; it changed how they learn. Training evolved from rote repetition to deliberate practice – structured repetition combined with reflection, feedback, and mental effort (Ericsson et al., 1993). Instructors began treating learning as a cognitive skill in itself: the ability to monitor, evaluate, and adapt one's performance.

This approach acknowledged that memory, reasoning, and judgment can be trained just as muscle memory can. A pilot isn't merely memorizing procedures; they're building mental models that predict how a situation will unfold. The stronger and more flexible those models, the more resilient the performance.

Behind these methods lie a few key principles that make training stick:

- **Neuroplasticity**: Repeated, effortful practice with feedback strengthens and refines neural pathways. Structured, goal-directed repetition changes performance more than unguided time in the simulator (Ericsson et al., 1993).
- **Retrieval and Variation:** Skills last when they're retrieved from memory and applied in varied contexts – mixing cues, changing scenarios, and challenging assumptions. This "desirable difficulty" builds flexible, use-anywhere competence (Roediger and Karpicke, 2006; Schmidt and Bjork, 1992).
- **Arousal and Attention:** Too little stress and we disengage; too much and the prefrontal cortex yields to reflex. Training should aim for the optimal middle zone and teach crews how to regulate back into it (Arnsten, 2009; Yerkes and Dodson, 1908).

- **Sleep and Consolidation:** Sleep after training consolidates learning; spacing sessions over days outperforms a single intense cram (Walker and Stickgold, 2006).
- **Culture as an Enabler:** A psychologically safe training environment keeps the brain's control systems online, allowing learners to explore, take feedback, and try again without fear (ICAO, 2018).

Together, these mechanisms explain why the best training programs don't just build skill; they reshape cognition.

CBTA AND THE COMPETENCY REVOLUTION

As aviation matured, a quiet revolution took shape. Training was no longer just about logging hours or repeating checklists. It became about cultivating judgment, communication, and adaptability under pressure – about shaping professionals who can think as well as act. This new mindset found its structure in **CBTA** (EASA, 2025; IATA, 2023). In this book, CBTA is used as a general international concept rather than referring to any specific regulatory framework.

CBTA is more than a change in paperwork. It is a fundamental redefinition of what it means to be "ready" to fly. Instead of focusing on task completion or box-ticking, it asks a deeper question: *Can this person apply knowledge and skill effectively in context, especially when conditions shift?*

From Proficiency to Competence

For decades, flight training had revolved around proficiency – a narrow, mechanical idea of correctness. You demonstrated a maneuver to standard, and that was enough. But competence demanded something richer. It connected technical skill to behavior, attitude, and cognition. It asked how people use what they know, not just whether they know it (EASA, 2025; ICAO, 2018).

CBTA recognized that aviation work is dynamic. Pilots, cabin crew, and engineers operate in environments where no checklist can cover every scenario. True safety depends on judgment in the gray areas – where standard operating procedures meet human decision-making.

The approach echoed decades of cognitive research. Expertise, as studies showed, is not simply repetition; it's the ability to interpret, adapt, and reframe problems when patterns break (Ericsson et al., 1993; Klein, 1999). CBTA turned this science into structure, linking each observable behavior to measurable evidence of competence.

In other words, it **operationalized thinking**.

The Structure of Competence

ICAO defines competence as a combination of knowledge, skills, and attitudes that together enable safe, effective performance (ICAO, 2016) Under the CBTA model, training outcomes are built around competency units – broad areas such as flight path management, communication, and SA – each described through **observable behaviors.**

These behaviors make the abstract visible. "Maintain situational awareness" becomes "verifies mode changes," "monitors cross-checks," or "verbalizes potential conflicts." Instructors can now assess what they see, not what they guess is happening internally.

This behavioral clarity supports **fair, transparent assessment**, reducing personal bias and aligning instruction across teams. It also reinforces the systems-thinking perspective: performance is judged in context, not isolation.

A crucial element of CBTA is its alignment with **Evidence-Based Training (EBT)**. Instead of building scenarios around worst-case imagination, programs use real-world data – to design training that mirrors actual operational threats (FAA, 2017; ICAO, 2018).

In this sense, CBTA isn't just a training framework. It's an adaptive safety system that learns through people.

The Instructor's New Role

CBTA also redefined what it means to teach. The instructor's authority shifted from examiner to facilitator, from judge to coach (IATA, 2023). Instead of asking, "Did they meet the standard?" instructors began asking, "How did they think through this?"

Feedback should be now reflective, not corrective. Debriefs turn into conversations about mental models, workload management, and communication choices. The old question "What went wrong?" is supposed to change into "What was your picture at that moment?"

This evolution required instructors to develop new competencies of their own – empathy, observation, and the ability to decode behavior without assuming intent. Teaching became less about perfection and more about insight (IFALPA, 2021).

In CBTA, a good instructor doesn't just mark performance; they interpret cognition. They look for the pattern behind the error, the reason behind the reaction. In short, they teach people how to think under pressure, not just how to comply.

This reorientation also changed the relationship between instructor and trainee. The cockpit is becoming less hierarchical, more collaborative – mirroring the CRM values that started the transformation decades earlier. Trust, openness, and psychological safety are not just ethical goals but operational necessities (Fabre et al., 2021).

Assessment as Learning

Traditional checks often measured the wrong thing: compliance over competence. CBTA reframed assessment as learning, not judgment.

Every observation – from decision-making to leadership – is rated against clear behavioral indicators. But the goal is not a number; it's a conversation. Scores are the entry point for feedback that deepens understanding and strengthens resilience (EASA, 2025).

This focus on **cognitive visibility** also helps identify latent potential. A pilot who hesitates to act may not lack knowledge; they may be overloading cognitively. The insight directs coaching, not punishment.

When these observations are recorded systematically, they become data for continuous improvement. Training notes, behavioral trends, and performance metrics feed into the learning system itself. If several crews show difficulty managing automation mode transitions, the curriculum evolves to address it.

This creates an adaptive ecosystem: the training organization learns from its own people. It's systems thinking applied to education – only now, the "system" is the training program. Integrated with proper SMS and EBT, this alignment turns data into foresight rather than hindsight, ensuring training stays relevant to real operational risk (ICAO, 2018; Kanki et al., 2019).

Looking Ahead – The Mind and Technology as the Next Safety Path

Fatigue, cognition, and automation bias are more than research topics; they mark the future of safety thinking. They push us to move beyond compliance toward comprehension – to ask not only *What went wrong?* but *What supports human performance when things go right?*

The future of human factors lies at this intersection: where biology meets behavior, and behavior meets technology. Artificial intelligence, adaptive training, and immersive simulation are already extending human capability – but they also challenge how we define trust, attention, and control. Research on **human-AI teaming in aviation** shows that the safest systems are those that respect human cognitive limits while keeping the operator in the loop (Kirwan, 2025). In the same time we need to keep highlighting transparency, shared intent, and human adaptability as essential to trustworthy automation (Khosrow-Pour, 2020).

Automation and AI are not distant concepts – they are shaping already how pilots train, communicate, and decide, themes explored further in *Chapter 10: Manual Control and Automation – Hands On or Hands Off?* and *Chapter 11: AI and Future Technologies in Human Performance.*

LEGACY MODELS AND THE NEED FOR EVOLUTION

Aviation grew up on models, also the ones rooted in human factors. They gave us pictures to think with and a common language to act on. **SHELL**, **Swiss Cheese**, or **TEM** (Hawkins 2017; ICAO, 2018; Maurino, 2005; Reason, 1990) weren't just academic ornaments; they changed how we brief, fly, debrief, and investigate – and they still do.

However, the world we fly in has shifted. Automation is more pervasive, data flow faster, and our understanding of cognition and variability is richer. So the question isn't "Are the classics wrong?" It's "Do they show enough of today's picture-and how do we extend them without losing their strengths?" (Hollnagel et al., 2015).

The Strength of the Classics

The idea that safety isn't just about the person in the cockpit but about the space around them began with Elwyn Edwards in the early 1970s. He proposed the original **SHELL** concept: Software, Hardware, Environment, and Liveware (SHEL), as a way to show that performance depends on how these parts fit together, not on one alone (Edwards, 1972). In later work, **Frank Hawkins** expanded the framework by explicitly emphasizing human–human interaction, often described as a second Liveware (L–L) interface (a development referred to as SHELL), and by introducing a building-block representation that placed the human at the center of the system (Hawkins, 1984; 2017).

This was one of the first times aviation tried to visualize safety as a web of relationships rather than a list of errors. It invited us to look at how pilots, machines, and environments interact, and how tiny misalignments between them can ripple into outcomes that no single person caused.

The Swiss Cheese Model, developed by James Reason in the 1990s, became one of the most recognized metaphors in safety thinking. It explained how accidents often occur not from a single error, but from a series of latent conditions and active failures aligning – "holes" in layers of defense (Reason, 1990). It allowed safety investigations to focus on system vulnerabilities rather than individual blame and has informed frameworks such as the Human Factors Analysis and Classification System (HFACS), helping investigators and regulators apply structured reasoning to human error analysis (Wiegmann and Shappell, 2003).

TEM was first developed in the late 1990s by the University of Texas Human Factors team (Helmreich, Klinect, and Merritt), and later adopted by ICAO in 2002. TEM carried systemic thinking straight into the flight deck. It framed performance around how crews detect, manage, and recover from threats and errors, embedding the spirit of CRM into everyday line operations (ICAO, 2018; Maurino, 2005).

These models reshaped how we train, debrief, and investigate. They remain brilliant conceptual tools. Their legacy lies in the shift they brought – from reactive to proactive, from blame to understanding.

The Gaps We Now See

At the same time, aviation has changed. Classic models were developed in an era that focused on error prevention and linear causality. But today's flight operations operate in complex, dynamic environments, where safety also depends on adaptation, creativity, resilience, and the ability to process uncertainty.

While the Swiss Cheese Model excels in post-incident analysis, it offers less insight into the everyday successes of human performance. It tends to emphasize what failed – less so what worked. Similarly, TEM is practical and operationally useful, but it may not capture the full picture of cognition, fatigue, or organizational influence (Metso et al., 2016; Wong et al., 2022).

Moreover, these legacy models were not built with emerging risks in mind – such as human–autonomy teaming, neurocognitive overload, or cybersecurity. Nor do they fully integrate the latest understanding from cognitive science, human–machine interface design, or behavioral variability under uncertainty.

None of this invalidates them. In fact, their strength is precisely that they remain relevant enough to serve as foundations. But it does suggest we now need models that include not only latent failures but also active strengths, not only threats but also adaptive capacity.

When Models Become Mental Traps

There is also a risk that once models become institutionalized, they shape the way we see the world. In investigations, we may try to fit data into the model, rather than adapting the model to fit the reality. In training, the visual simplicity of Swiss Cheese or SHELL may lead to overconfidence in prediction and control (Wiegmann and Shappell, 2003; Wong et al., 2022). Over time, what begins as a helpful abstraction can quietly harden into a lens through which events are interpreted, sometimes retrospectively, sometimes uncritically.

This isn't a flaw of the models themselves – it's a function of how they are used. As powerful as they are, they are not complete explanations. They are starting points for thinking, not final answers.

Used rigidly, models may limit our ability to ask better questions. Used flexibly, they provide essential scaffolding for understanding.

Models as Evolving Tools, Not Static Truths

Models help us explain, teach, and explore, not decree truth. We still need SHELL to interrogate interfaces and Swiss Cheese to think in layers, with TEM to connect this to daily work (Hawkins,2017; ICAO, 2018; Maurino, 2005; Reason, 1990).

But models must evolve as our understanding deepens. Today, we know more about cognition, neurophysiology, systems theory, and emotional intelligence than we did in the 1990s. We also work in more automated, data-rich, and socially complex aviation environments. That doesn't make older models wrong – it simply means that the lens we use to view safety must grow wider and more dynamic.

That's why, in my work, I began introducing an additional conceptual tool. It was never intended to replace the existing ones, but rather to sit alongside them, helping to visualize some of the relationships and influences we now know are critical. Over time, this model became known as the **SPINAL-CORD model**.

In one view, it shows the human inside the system and the system around the human. It reminds us that safety isn't produced by people or procedures alone – it's co-produced at the interface. We will use it as a visual anchor for training, debrief, and investigation throughout the book.

INTRODUCING THE SPINAL-CORD MODEL: A MODERN SAFETY INTEGRATION HEURISTIC

Understanding safety in aviation today requires more than counting errors or enforcing compliance. It demands a lens that integrates systems, people, and context. In that spirit, the **SPINAL-CORD model** was developed, not as a replacement for established frameworks like Swiss Cheese or SHELL, but as a complementary visual heuristic for teaching, reflection, and discussion.

Originally designed to help students, crews, and managers grasp the complexity of human performance, SPINAL-CORD blends the legacy of human factors with modern science. It invites us to see pilots not as the final line of defense,

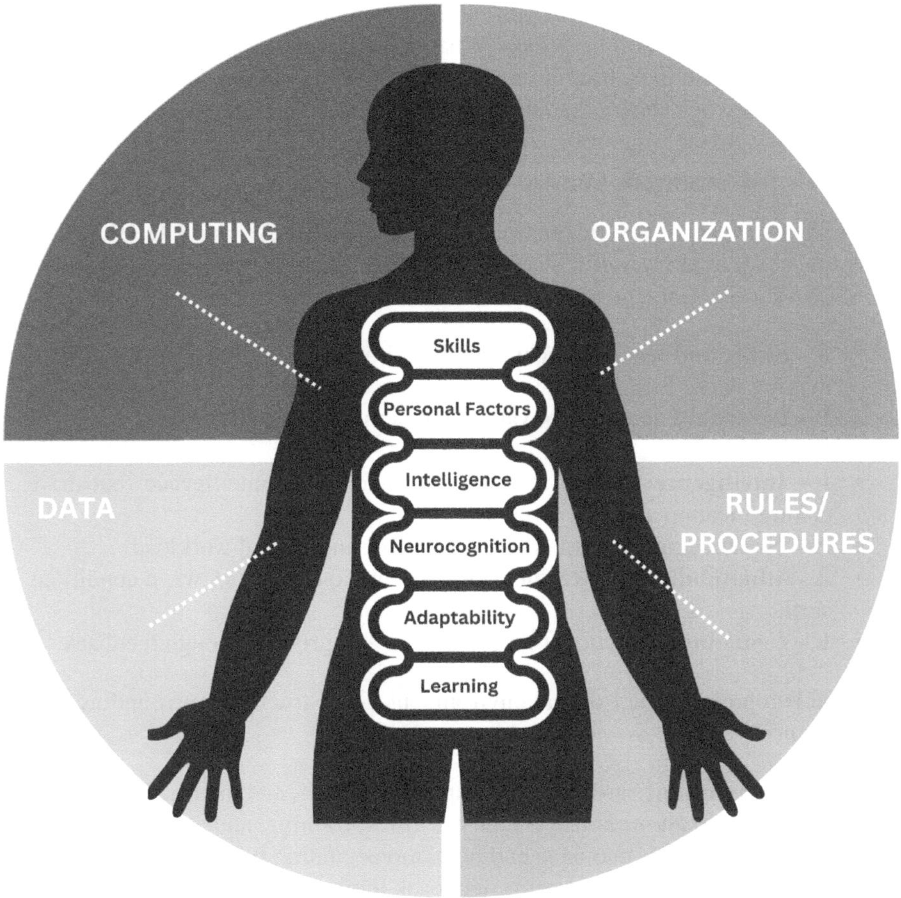

FIGURE 1.3 SPINAL-CORD model – system outside, human inside.

but as integral parts of a living system where safety is continuously co-produced (Figure 1.3).

What the Model Aims to Do

The SPINAL-CORD model is not a rulebook – it's a thinking tool. Drawing on established safety theory and cognitive science, it visualizes how multiple human and systemic layers interact in real operations.

While Swiss Cheese illustrates how defenses can fail and SHELL categorizes human–system interfaces, SPINAL-CORD shows the dynamic balance between:

- Human capabilities and constraints
- Organizational structures and tools
- Cognitive strain and adaptive strategies

This perspective helps instructors and investigators move beyond "who did what wrong" to the more useful question: "What was shaping this behavior?" It also provides a visual anchor in training and debriefings where complex interactions are easier to grasp when shown than described.

Layers of the Model: A Systems – Human Continuum

The model consists of two interacting components: **SPINAL** and **CORD**, each representing a different layer of the safety landscape.

SPINAL represents the human and behavioral core:

- **S – Skills and Competence:** operational proficiency, CRM, and decision-making under load.
- **P – Personal Factors:** motivation, temperament, traits, attitudes, stress, or fatigue.
- **I – Intelligence:** cognitive, emotional, and social intelligence that drive communication and adaptability.
- **N – Neurocognition:** attention, biases, SA, and mental workload.
- **A –Adaptability and Resilience:** flexibility and recovery when conditions shift.
- **L – Learning:** reflection, self-correction, and growth through feedback.

CORD represents the external environment—the structures surrounding human performance:

- **C – Computing:** hardware, software, automation, and interface design.
- **O – Organization:** management structures, training, and culture.
- **R – Rules:** SOPs, manuals, and regulations guiding behavior.
- **D – Data:** monitoring systems, feedback loops, and safety intelligence.

SPINAL represents the living human system – the dynamic interaction of cognition, emotion, and behavior. CORD depicts the engineered system – the design, rules, and data that shape performance. Together, they describe how safety emerges from interaction, not isolation.

Adaptive Barriers, Cognitive Filters

A core insight of SPINAL-CORD is that humans are not simply points of failure or last defenses – they are filters, sensors, and adaptive barriers. Cognitive overload, emotional strain, or mental shortcuts can either amplify or block risk.

The model prompts us to ask:

- *Was a lapse due to lack of skill – or a system design that made skill harder to apply?*
- *Are procedures cognitively aligned with how people actually perceive and act?*
- *How did learning, feedback, or resilience transform a near-miss into a safe recovery?*

This framing shifts focus from error prevention to performance understanding. It asks *how conditions shaped perception and choice*, not merely what went wrong afterward.

Training, design, and leadership therefore become co-responsible for managing both sides of safety: **system barriers** (clear SOPs, automation transparency) and **human adaptability** (attention, stress regulation, decision-making).

Using the Model in Practice

The SPINAL-CORD model is already used in flight school instruction, safety workshops, and debriefs to help translate theory into operational language. Its flexibility allows it to support:

- **Instructor and Examiner Training:** mapping observed behaviors to SPINAL elements to explore what shaped performance.
- **Incident Analysis:** identifying system and human layers that interacted in an event, rather than isolating a single cause.
- **Safety Briefings and Campaigns:** framing resilience and human–system balance as ongoing processes, not fixed traits.

Because of its simplicity, the model encourages curiosity. Instructors often use it as a backdrop for reflection sessions – asking students to locate an observed behavior within SPINAL (internal factors) or CORD (external influences). The exercise often reveals patterns that conventional checklists overlook: cultural cues, cognitive overload, or procedural friction points.

In practice, SPINAL-CORD serves three complementary purposes:

1. **As a map** – clarifying how human and system factors align or conflict.
2. **As a mirror** – helping individuals and organizations see where strain accumulates.
3. **As a bridge** – linking human factors science to operational dialogue in a non-technical way.

WHY SPINAL-CORD MATTERS

SPINAL-CORD shifts focus from preventing failure to understanding what keeps performance strong. It integrates modern insights from cognition, resilience, and adaptation – areas older frameworks rarely captured. By pairing SPINAL (the adaptable human) with CORD (the structured system), it visualizes co-production: safety emerging not from top-down control or individual skill alone, but from their constant interaction.

The model bridges classical and contemporary human factors. Swiss Cheese still explains how defenses align; SHELL clarifies interfaces; SPINAL-CORD extends them to include cognition, emotional regulation, and feedback. It connects traditional safety logic with the living dynamics of human performance.

This evolution mirrors aviation itself – from static procedures to adaptive competence, from error counting to performance learning, and from isolated operators to interdependent ecosystems. It supports the move from reactive analysis to proactive learning, helping organizations explore not only why incidents occur but why success happens despite complexity.

Ultimately, SPINAL-CORD reflects this book's broader purpose: to bridge tradition and innovation, honoring the legacy of human factors while giving language to today's challenges. It reminds us that every aviation professional is not a weak link, but a living part of the safety system – thinking, adapting, and balancing limitation with ingenuity.

TAKEAWAYS: STARTING OUR HUMAN FACTORS JOURNEY

For Pilots and Cabin Crew

- Take time to understand how system design and culture shape your performance – not just your own skills.
- Reflect regularly on your adaptability and what helps you recover from high workload or change.
- Use visual models (like SPINAL-CORD or others) to mentally rehearse your interaction with the system around you.

Reflective Question: *What personal trait or system factor has most influenced your performance during a challenging flight?*

For Instructors, Examiners, and Leaders

- When debriefing or instructing, help students recognize how the system (tech, rules, culture) interacts with behavior.
- Use models to teach – not as rigid structures, but as tools for thinking, discussing, and exploring.
- Encourage your learners to reflect not only on what went wrong, but what helped them succeed.

Reflective Question: *How can you better support your trainees in developing adaptability and learning beyond the checklist?*

For Operational Teams (Different Levels)

- Design procedures and training frameworks that acknowledge the full system around the human, not just individual competence.
- Revisit older safety models, but enrich them with insights from cognitive science, systems theory, and data.
- Consider adopting new visual tools to support briefing, safety culture, and performance analysis.

Reflective Question: *Are your safety interventions focused more on preventing error-or also on enabling human strengths?*

REFERENCES

Air Accident Investigation Sector (AAIS) (2020) *Final Report: Boeing 777-31H, A6-EMW, Emirates Flight 521, Dubai International Airport, 3 August 2016.* Abu Dhabi: General Civil Aviation Authority (GCAA), United Arab Emirates. Available at: https://www.icao.int/sites/default/files/safety/airnavigation/AIG/Documents/Safety-Recommendations-to-ICAO/Final-Reports/UAE_B777_A6-EMW_3Aug2016.pdf

Arnsten, A. F., Raskind, M. A., Taylor, F. B. and Connor, D. F. (2015) 'The effects of stress exposure on prefrontal cortex: Translating basic research into successful treatments for post-traumatic stress disorder', *Neurobiology of Stress*, 1, pp. 89–99. https://doi.org/10.1016/j.ynstr.2014.10.002

Arnsten, A. F. T. (2009) 'Stress signalling pathways that impair prefrontal cortex structure and function', *Nature Reviews Neuroscience*, 10(6), pp. 410–422. https://doi.org/10.1038/nrn2648

Cooper, G. E., White, M. D. and Lauber, J. K. (1980) *Resource Management on the Flightdeck: Proceedings of a NASA/Industry Workshop.* NASA CP-2120. Moffett Field, CA: NASA Ames Research Center. https://ntrs.nasa.gov/api/citations/19800013796/downloads/19800013796.pdf

Dawson, D. and Reid, K. (1997) 'Fatigue, alcohol and performance impairment', *Nature*, 388(6639), 235. https://doi.org/10.1038/40775

Dekker, S. (2002). *The Field Guide to Human Error Investigations.* 1st ed. Abingdon: Routledge. https://doi.org/10.4324/9781315202778

Dekker, S. (2012). *Just Culture: Balancing Safety and Accountability.* 2nd ed. Boca Raton, FL: CRC Press. https://doi.org/10.4324/9781315251271

Dekker, S. (2014) *The Field Guide to Understanding 'Human Error'.* 3rd ed. Boca Raton, FL: CRC Press. https://doi.org/10.1201/9781317031833

European Union Aviation Safety Agency (EASA) (2025) *Evidence-Based Training (EBT) Manual*, Version 2.3. Cologne: EASA. Available at: https://www.easa.europa.eu/en/domains/aircrew-and-medical/evidence-based-training-ebt

Edwards, E. (1972) *'Man and machine: Systems for safety, Proceedings of the British Airline Pilots Association Technical Symposium*, London: BALPA.

Endsley, M. R. (1995) 'Toward a theory of situation awareness in dynamic systems', *Human Factors*, 37(1), pp. 32–64. https://doi.org/10.1518/001872095779049543

Endsley, M. R. (2018). 'Expertise and situation awareness', in K. A. Ericsson, R. R. Hoffman, A. Kozbelt, A. M. Williams (Eds.), *The Cambridge Handbook of Expertise and Expert Performance* (2nd ed., pp. 714–741). Cambridge, MA: Cambridge University Press. https://doi.org/10.1017/9781316480748.037

Ericsson, K. A., Krampe, R. T. and Tesch-Römer, C. (1993) 'The role of deliberate practice in the acquisition of expert performance', *Psychological Review*, 100(3), pp. 363–406. https://doi.org/10.1037/0033-295X.100.3.363

Fabre, E. F., Matton, N., Beltran, F., Baragona, V., Cuny, C., Imbert, J.-P., Voivret, S., Van Der Henst, J.-B. and Causse, M. (2021) 'Hierarchy in the cockpit: How captains influence the decision-making of young and inexperienced first officers', *Safety Science*, 140, p. 105536. https://doi.org/10.1016/j.ssci.2021.105536

Federal Aviation Administration (FAA) (2017) *Advisory Circular 120-54A: Advanced Qualification Program* - (Change 1). Washington, DC: Federal Aviation Administration. Available at: https://www.faa.gov/documentLibrary/media/Advisory_Circular/AC_120-54A_CHG_1.pdf

Hawkins, F. H. (1984) *Human Factors education in European air transport operations In: Breakdown in Human Adaptation to Stress. Towards a multidisciplinary approach, Vol. 1, for the Commission of the European Communities.* The Hague: Martinus Nijhoff.

Hawkins, F. H. (2017) *Human Factors in Flight*. 2nd ed. London: Routledge.
Helmreich, R. L. and Merritt, A. C. (2001) *Culture at Work in Aviation and Medicine: National, Organizational and Professional Influences*. 1st ed. Abingdon: Routledge. https://doi.org/10.4324/9781315258690
Hollnagel, E. (2014) *Safety-I and Safety-II: The Past and Future of Safety Management*. Farnham: Ashgate. https://doi.org/10.1201/9781315607511
Hollnagel, E., Wears, R. L. and Braithwaite, J. (2015) *From Safety-I to Safety-II: A White Paper*. The Resilient Health Care Net. https://www.england.nhs.uk/signuptosafety/wp-content/uploads/sites/16/2015/10/safety-1-safety-2-whte-papr.pdf
Hollnagel, E. and Woods, D. D. (2005). *Joint Cognitive Systems: Foundations of Cognitive Systems Engineering*. 1st ed. Boca Raton, FL: CRC Press. https://doi.org/10.1201/9781420038194
International Air Transport Association (IATA) (2023) *Competency-Based Training and Assessment (CBTA) Expansion within the Aviation System*. Montreal: IATA. Available at: https://www.iata.org/contentassets/c0f61fc821dc4f62bb6441d7abedb076/cbta-expansion-within-the-aviation-system.pdf
International Civil Aviation Organization (ICAO) (1978) *Accident Investigation Report: Collision of KLM Flight 4805 and Pan Am Flight 1736, Tenerife, 27 March 1977*. Montreal: ICAO. Available at: https://www.faasafety.gov/files/gslac/courses/content/232/1081/finaldutchreport.pdf
International Civil Aviation Organization (ICAO) (1984) *Human Factors in Civil Aviation Operations*. Circular 153-AN/56. Montreal: ICAO.
International Civil Aviation Organization (ICAO) (2016) *Training Development Guide: Competency-Based Training Methodology*, Doc 9941. Montreal: ICAO. Available at: https://store.icao.int/en/training-development-guide-competency-based-training-methodology-doc-9941
International Civil Aviation Organization (ICAO) (2020) *Manual for the Oversight of Fatigue Management Approaches*, Doc 9966, Second edition, version 2 (revised). Montreal: ICAO. Available at: https://elibrary.icao.int/product/234625 / https://store.icao.int/en/manual-for-the-oversight-of-fatigue-management-approaches-doc-9966
International Civil Aviation Organization (ICAO) (2018) *Safety Management Manual (SMM). 4th ed. Doc 9859*. Montreal. Available at: https://store.icao.int/en/safety-management-manual-doc-9859/ https://elibrary.icao.int/product/229751
International Civil Aviation Organization (ICAO) (2023) *Doc 10004 – Global Aviation Safety Plan (GASP) 2023–2025*. Montreal. Available at: https://www.icao.int/sites/default/files/safety/GASP/Documents/10004_en.pdf
International Federation of Air Line Pilots' Associations (IFALPA) (2021) *Guidance Material and Best Practices for Instructor and Evaluator Training*. 2nd ed. Montreal: IFALPA. Available at: https://www.ifalpa.org/media/3630/guidance-material-and-best-practices-for-instructor-and-evaluator-training.pdf
Kanki, B. G., Anca, J. and Chidester, T. R. (2019) *Crew Resource Management*. 3rd ed. London: Academic Press.
Khosrow-Pour, M. (Ed.) (2020) *Handbook of Research on Artificial Intelligence in Aviation and Aerospace*. Hershey, PA: IGI Global.
Kirwan, B. (2025). *Human factors requirements for human-AI teaming in aviation*. Future Transportation, 5(2), 42. https://doi.org/10.3390/futuretransp5020042
Klein, G. (1999) *Sources of Power: How People Make Decisions*. Cambridge, MA: MIT Press.
Martin, W., Murray, P., Bates, P. and Lee, S. Y. (2016) 'A flight simulator study of the impairment effects of startle on pilots during unexpected critical events', *Aviation Psychology and Applied Human Factors*, 6(1), pp. 24–32. https://doi.org/10.1027/2192-0923/a000092
Maurino, D. E. (2005) 'Threat and error management (TEM) and the challenge of aviation safety', *The International Journal of Aviation Psychology*, 15(1), pp. 1–6. https://doi.org/10.1207/s15327108ijap1501_1

Metso, L., Marttonnen, S., Thenent, N. and Newnes, L. (2016) 'Adapting the SHEL model in investigating industrial maintenance', *Journal of Quality in Maintenance Engineering*, 22(1), pp. 62–80. https://doi.org/10.1108/JQME-12-2014-0059

Miller, G. A. (1956) 'The magical number seven, plus or minus two: Some limits on our capacity for processing information', *Psychological Review*, 63(2), pp. 81–97. https://doi.org/10.1037/h0043158

National Transportation Safety Board (NTSB) (2024) *Aviation Investigation Report AIR-24-02: Runway Incursion and Overflight, Southwest Airlines Flight 708/FedEx Flight 1432, Austin, Texas, February 4, 2023*. Washington, DC: NTSB. Available at: https://www.ntsb.gov/investigations/AccidentReports/Reports/AIR2402.pdf

Parasuraman, R. and Riley, V. A. (1997) 'Humans and automation: Use, misuse, disuse, abuse', *Human Factors*, 39(2), pp. 230–253. https://doi.org/10.1518/001872097778543886

Reason, J. (1990) *Human Error*. Cambridge, MA: Cambridge University Press.

Reason, J. (1997) *Managing the Risks of Organizational Accidents*. Aldershot: Ashgate.

Roediger, H. L. and Karpicke, J. D. (2006) 'Test-enhanced learning: Taking memory tests improves long-term retention', *Psychological Science*, 17(3), pp. 249–255. https://doi.org/10.1111/j.1467-9280.2006.01693.x

Ross, L. (1977) 'The intuitive psychologist and his shortcomings: Distortions in the attribution process', in L. Berkowitz (Ed.), *Advances in Experimental Social Psychology*, 10, 173–220. New York: Academic Press.

Schmidt, R. A. and Bjork, R. A. (1992) 'New conceptualizations of practice: Common principles in three paradigms suggest new concepts for training', *Psychological Science*, 3(4), pp. 207–217. https://doi.org/10.1111/j.1467-9280.1992.tb00029.x

Swauger, T. (2023) *The Illusion of Control: Decision Making under Pressure in Aviation Operations*. Washington, DC: Aviation Safety Institute Press.

UK Civil Aviation Authority (CAA) (2002) *CAP 719: Fundamental Human Factors Concepts in Civil Aviation*. London: Civil Aviation Authority. Available at: https://publicapps.caa.co.uk/modalapplication.aspx?appid=11&mode=detail&id=226

Van Dongen, H. P. A., Caldwell, J. A. and Caldwell, J. L. (2006) 'Investigating systematic individual differences in sleep-deprived performance on a high-fidelity flight simulator', *Behavior Research Methods*, 38, pp. 333–343. https://doi.org/10.3758/BF03192785

Walker, M. P. and Stickgold, R. (2006) 'Sleep, memory and plasticity', *Annual Review of Psychology*, 57, pp. 139–166. https://doi.org/10.1146/annurev.psych.56.091103.070307

Weick, K. E. (1990) 'The vulnerable system: An analysis of the Tenerife air disaster', *Journal of Management*, 16(3), pp. 571–593. https://doi.org/10.1177/014920639001600304

Wickens, C. D. (2008) 'Multiple resources and mental workload', *Human Factors*, 50(3), pp. 449–455. https://doi.org/10.1518/001872008X288394

Wickens, C. D., Liang, Y., Goh, J. and Goh, S. C. (2022) *Human Factors in Flight Deck Design and Operation*. Boca Raton, FL: CRC Press.

Wiegmann, D. A. and Shappell, S. A. (2003) *A Human Error Approach to Aviation Accident Analysis: The Human Factors Analysis and Classification System*. Aldershot: Ashgate.

Wiener, E. L. and Nagel, D. C. (2010) *Human Factors in Aviation*. 2nd ed. San Diego, CA: Academic Press.

Wong, B. L. W., Blandford, A., McIlroy, R. C. and Stedmon, A. W. (2022) 'A review of accident models and incident analysis techniques for complex sociotechnical systems', *Safety Science*, 148, p. 105623.

Yerkes, R. M. and Dodson, J. D. (1908) 'The relation of strength of stimulus to rapidity of habit-formation', *Journal of Comparative Neurology and Psychology*, 18(5), pp. 459–482. https://doi.org/10.1002/cne.920180503

2 Safety Culture
Beyond Compliance – A Safety Mindset

CASE STUDY: A TEST OF SAFETY CULTURE – INSIDE SOUTHWEST 1380 (2018)

On April 17, 2018, **Southwest Airlines Flight 1380** departed New York for Dallas with 144 passengers and five crew members. At 32,000 feet, the calm routine of cruise flight shattered when the left engine suffered an uncontained failure. Debris punctured the fuselage, breaking a window and causing rapid depressurization. Oxygen masks dropped; passengers screamed; one was partially ejected.

Inside the cockpit, Captain Tammie Jo Shults, a former Navy fighter pilot, and First Officer Darren Ellisor worked quickly to regain control. The aircraft rolled and shuddered, alarms overlapped, and communications crackled, yet their tone remained calm. They declared an emergency, began a controlled descent, and coordinated every action through short, clear exchanges. In the cabin, flight attendants moved through fear and confusion with the same professionalism: coaching passengers through oxygen use, brace positions, and preparation for a landing no one had expected. In post-event reflections, the captain acknowledged that the response was not technically perfect (NTSB, 2019; Shults, 2019), a reality that reflects the conditions under which the crew were operating rather than a competence issue. The event unfolded under extreme startle, noise, and time pressure, where priorities shifted rapidly toward control and survival.

Philadelphia was not the nearest airport, but it offered the longest runway and the best support. The crippled 737 landed safely minutes later. One passenger died (the partially ejected one), but all 148 others survived.

The National Transportation Safety Board (NTSB) later praised the crew's exceptional airmanship and noted Southwest's immediate response: inspections across the fleet, new training modules, and transparent data sharing with the industry. Yet the deeper lesson was taught that day. The same engine type had shown warning signs before, but inspection intervals had stretched, and roles between airline, manufacturer, and regulator had blurred (NTSB, 2019). This case is therefore not presented as an example of flawless execution or perfect system, but as an illustration of how safety can often be salvaged through adaptive performance. Importantly, the broader lesson extends beyond the cockpit: it highlights the role of cabin crew, ground services, emergency responders, and the organization itself in supporting recovery.

DOI: 10.1201/9781003515418-2

INTRODUCTION

Southwest 1380 is more than a story of technical failure, it's a window into how culture reveals itself under pressure. Aviation safety has never rested on a single checklist or definition. It is layered, and those layers often become visible only when something cracks (Reason, 1997).

Checklists, Standard Operating Procedures (SOPs), and regulations are essential, but they are scaffolding, not the foundation. What sustains aviation safety is culture – the shared beliefs and behaviors that guide choices when supervision is absent, when procedures end, and when the decision sits in a gray area.

Safety culture goes beyond compliance. It reflects how risk is managed in the everyday details: how leaders speak about safety when profits are tight, how engineers raise concerns in the hangar, how a controller frames workload during a shift, or how a captain models decision-making on the line. It is lived, not laminated (Dekker, 2014; ICAO, 2018; Reason, 1997).

In this chapter, we will explore safety culture not as a static theory but as a living ecosystem – visible in the choices of frontline staff, shaped by leadership behavior, embedded in organizational routines, and influenced by industry-wide dynamics. The journey to a genuine safety mindset demands more than audits or slogans. It requires a conscious, collective shift from reactive compliance to proactive responsibility.

We will consider how culture manifests at different layers – within individuals, inside teams, across organizations, and through the wider industry. We will reflect on how it evolves, how it is tested in moments of crisis, and how subtle warning signs of drift can be spotted before they lead to consequences. And perhaps most importantly, we will see how small, consistent actions (like those taken by the Southwest crew that morning) can either reinforce or erode the cultural foundations on which aviation safety depends.

LEVELS OF SAFETY CULTURE – THE DIMENSIONS WE CAN OBSERVE

The phrase ***safety culture*** is common in aviation, yet from a human factors' view it's often misunderstood. The case of Southwest 1380 reminds us that safety is multilayered; it lives in individual attitudes and habits, in how teams coordinate and communicate under pressure, and in the values and leadership behaviors that define how an organization manages risk (Reason, 1997).

Rather than offering another copy-and-paste definition, it's more useful to understand safety culture by exploring these layers and how they interact and change over time. None of them can stand alone. A strong culture depends on alignment across all three: individuals with the right mindset, cohesive teams, and organizations that encourage awareness, openness, and continuous improvement.

Safety Culture on the Individual Level

At the individual level, safety culture is about personal attitudes, perceptions, and behaviors. It's where psychology meets operations. Pilots, controllers, cabin crew, engineers – even managers and executives – carry their own safety mindset into every shift. Traits such as conscientiousness, emotional stability, and risk perception all play a role in whether someone speaks up, follows through, or notices the small anomaly that could snowball into an incident (O'Connor et al., 2011).

Helmreich and Merritt (2001) showed that when pilots hold strong internal safety attitudes, they don't just comply with rules – they anticipate threats, recognize weak signals, and act early. It isn't about reciting checklists; it's about mindset.

Flight 1380 showed this vividly. Captain Shults' composure and clear decision-making, paired with First Officer Ellisor's proactive support and communication, were not products of chance. They were the outward expression of their individual-level safety culture – resilience, professional discipline, and an unwavering sense of responsibility in crisis.

Safety Culture on the Team Level

Teams are where individual mindsets either strengthen or cancel each other out. At this level, safety culture means communication, trust, coordination, and the ability to hold a shared mental picture of the unfolding situation. When crews practice effective Crew Resource Management (CRM), they don't just move through tasks – they shape the flow together. They back each other up, exchange information, and balance workload in ways that prevent small issues from escalating (Salas et al., 2015).

Southwest 1380 again brought this into focus. In the cockpit, Shults and Ellisor worked not only at the level of aircraft control, but also in how they coordinated as a team– trading roles without confusion, anticipating each other's needs, and never letting emotion cloud their focus. In the cabin, flight attendants modeled the same professionalism, guiding passengers through fear and confusion. Even passengers themselves, under crew direction, helped manage the emergency (NTSB, 2019). That was safety culture at the team level, not theory, but real coordination under fire.

Moments of acute stress or startle act like stress tests for the system, stripping away routines and exposing what is truly embedded: habits, attitudes, and unspoken norms. Under sudden pressure, people do not rise to the level of written procedures; they fall back on what has been practised, tolerated, or silently learned. Gaps in safety culture often emerge not through dramatic rule-breaking, but through subtler behaviors – a concern not voiced, a change not communicated, an assumption left unchallenged. It is in these compressed, uncomfortable moments that the strength or fragility of a safety culture becomes unmistakably clear (Reason, 1997; CAA, 2002). However, when time, workload, and uncertainty converge in real operations, these dynamics become visible in ways that theory alone cannot fully capture.

Yeti Airlines Flight 691 (NYT691) was an ATR 72-212A (ATR 72-500), operating a scheduled flight from Kathmandu to Pokhara, Nepal, on 15 January 2023. During a high-workload visual approach to Runway 12 at the newly opened Pokhara International Airport, the aircraft experienced a sudden loss of effective thrust and rapidly deteriorating energy state, impacting terrain short of the runway.

The investigation found no evidence of a pre-impact mechanical failure and concluded that the most probable cause was the inadvertent selection of both propeller condition levers to the feather position, leading to dual propeller feathering, loss of thrust, aerodynamic stall, and collision with terrain. The report also highlighted a tight visual circuit with limited margin, elevated task demand, incomplete cross-checking during configuration changes, and breakdowns in SOP discipline and sterile cockpit routines (Aircraft Accident Investigation Commission of Nepal [AAIC], 2023).

Used here, this case is not a commentary on character or competence. Rather, it illustrates how safety culture at the team level becomes visible under stress: how crews share (or fail to share) a common mental model, how actions are verbalized and acknowledged, how configuration changes are cross-checked, and how deviations are challenged in real time. Startle, attentional narrowing, and task saturation do not create unsafe behavior on their own; they amplify existing habits and norms (CAA, 2002; EASA, 2025). When time and altitude remove the usual buffer, even small weaknesses in coordination, monitoring, or challenge-response can quickly shape the outcome.

This type of event reminds us that safety culture at the team level is not tested during stable, predictable operations. It is tested when something unexpected interrupts the flow, and the team must recover shared understanding under pressure.

Safety Culture on the Organizational Level

At the **organizational level**, safety culture takes shape through leadership, policies, training, and the systems that define how risk is managed day-to-day. Organizations with strong cultures do more than write procedures – they set expectations, allocate resources to safety, and encourage reporting without fear of reprisal (Dekker, 2012). They also actively manage fatigue, workload, and information flow, recognizing that operational safety is fragile when these areas are neglected.

Leadership behavior matters enormously. Zohar (2010) showed that when leaders consistently model safety as a priority, it cascades downward and becomes part of how teams and individuals act. This alignment creates a climate where vigilance is the norm, not the exception.

Southwest's response to Flight 1380 reflected this organizational layer. They quickly launched fleet-wide inspections, enhanced training, and shared lessons across the industry (NTSB, 2019). That willingness to act transparently and proactively reinforced their culture, signaling that safety improvement was not a slogan but a commitment embedded in the airline's DNA.

THE EVOLUTION OF SAFETY CULTURE

Safety culture did not appear overnight. It has grown slowly, shaped by hard lessons, tragic accidents, and the evolving understanding of how humans behave in complex, high-stakes systems. To grasp what safety culture means today, it helps to trace the journey – from an early focus on machines and mechanical failure to a view that now recognizes people as the central resource for safety in uncertainty.

From Mechanical Reliability to Human Complexity

The story of aviation safety began with a mechanical mindset. For decades, progress was measured by reliability: stronger parts, better engines, and stricter procedures. When something failed, the fix was usually more engineering or more rules. That worked early on, but incidents soon exposed its limits. Aircraft could be mechanically sound, but accidents still were happening, increasingly caused not by broken hardware but by the interaction of humans, teams, and organizations (Hollnagel, 2009; Wiegmann et al., 2002).

This realization opened the door to new safety paradigms – what Erik Hollnagel later framed as *Safety-I* and *Safety-II* (Hollnagel, 2014).

- **Safety-I** is the traditional lens: identify errors after they happen, tighten procedures, and build defenses so they don't happen again. It treats humans largely as risks – unreliable links in the chain who need oversight (Reason, 1997). This approach gave aviation its structure: checklists, reporting systems, and error-prevention strategies. But it is reactive. It works for repeatable, predictable scenarios, yet it struggles with the complexity and variability of modern operations.
- **Safety-II** flips the perspective. Instead of asking why things fail, it asks how they succeed. It sees human performance as adaptive, resilient, and essential in making the system safe. The focus shifts from avoiding error to cultivating capacity – training people to adapt, improvise, and recover when conditions change. Success, not just failure, becomes a source of learning (Hollnagel, 2014).

A newer concept, sometimes referred to as **Safety-III** (Leveson, 2020a), goes a step further. It emphasizes proactive system design and modeling to identify unsafe interactions before they arise. Instead of relying on post-event analysis or individual adaptability, Safety-III seeks to embed learning, transparency, and digital foresight directly into operations (Aven, 2022). It bridges engineering, human factors, and data science – showing that future safety will depend not only on human awareness but on how well systems are designed to anticipate, communicate, and prevent risk from forming in the first place.

These perspectives are not in competition so much as they are stages in an ongoing evolution. Safety-I gave aviation its backbone of standardization. Safety-II brought an appreciation of human adaptability. Safety-III hints at what may come next: system-wide thinking that mobilizes technology and design as part of culture. Together, they show how far the industry has come from its mechanical origins, and how far it may yet go.

Challenges in Adopting the Safety-II Mindset

The shift did not come easily. For generations, aviation professionals were trained to think in Safety-I terms. Instructors, examiners, and supervisors reinforced the message that the safest pilot was the one who avoided error at all costs (Kahneman, 2011).

The human brain itself reinforces this pattern. From childhood, we learn through error: touch a hot surface, feel pain, avoid it next time. Our neural wiring makes us vigilant for mistakes – it is easier to spot what went wrong than to analyze what kept things right. Success is invisible. Error is obvious (Leveson, 2020b).

That bias makes Safety-II harder to adopt. Identifying errors and learning from them remains essential, but if error is all we look for, we create blind spots. In complex environments, root causes are rarely clear, and yesterday's fix may not prevent tomorrow's accident. Worse, a fixation on avoiding mistakes can raise stress and reduce flexibility. People become rigid when what's needed is adaptation (Dekker, 2019).

Safety-II reframes the challenge. Instead of trying to stamp out every error, it asks: how do we keep the operation safe even when errors occur? That means training not just in checklists, but in competencies like resilience, communication, and improvisation. It means empowering individuals to share good practices, not only to report failures (Hollnagel, 2014). And it means recognizing that success often comes from the small, nearly invisible adjustments people make to keep the system on track.

Aviation will never eliminate error. But by valuing what goes right, Safety-II helps us build capacity to absorb shocks, recover quickly, and continue operating safely when conditions deviate from the script. Safety-III extends that thinking even further – asking how we might design entire systems so that resilience is built in, rather than left to chance.

SAFETY CULTURE IN PRACTICE: WHAT WE OBSERVE VS. WHAT WE INTEND

Even the best-written safety policies can fail if they aren't mirrored in practice. A strong safety culture doesn't live in binders or posters – it appears in the small, everyday choices people make when no one is watching. Do pilots speak up during briefings or let silence linger? Are engineers heard when they raise concerns? Do leaders show vulnerability or only authority? The gap between what's written and what's lived is where culture reveals itself.

> One story that exposed this gap was **Dana Air Flight 992** in Nigeria in 2012. The McDonnell Douglas MD-83 suffered dual engine failure and crashed into a residential area of Lagos, killing 153 people on board and several on the ground. One engine had been repeatedly flagged for low oil pressure, but no corrective action followed. Technicians later admitted feeling uneasy signing off the aircraft, yet pressure and assumptions about "someone else" taking responsibility kept them silent. On paper, procedures existed; in practice, silence prevailed. The failure was not only mechanical but cultural: a system where speaking up felt futile (AIB Nigeria, 2017).

This gap between manuals and reality lies at the heart of safety culture. When people don't feel psychologically safe to challenge, or believe their concerns won't matter, the system becomes fragile, so compliant on the surface but brittle underneath. Culture reveals itself not in crises but in routine: a skipped checklist, a quiet hesitation, the tone of feedback after a near miss. These micro-behaviors decide whether culture strengthens or erodes. It's not the slogan above the hangar door – it's the eye contact, the honesty, the listening.

Normalization of Drift

One of the most dangerous aspects of safety culture is that it rarely collapses in a single moment, it erodes slowly. What begins as a small deviation, like a missed briefing detail, a deferred defect, a shortcut to save time, gradually becomes the new normal. This is what Diane Vaughan identified in her study of the Challenger disaster: the **normalization of deviance** (Vaughan, 1996).

In aviation, this drift often feels harmless. A pilot skips a cross-check; nothing happens. An engineer signs off a snag under pressure; the aircraft flies fine. The absence of immediate consequences creates a feedback loop, each uneventful outcome reinforcing that the deviation is safe. Over time, the exception becomes the routine.

Psychology explains why this drift is so insidious. **Cognitive dissonance** allows people to justify bending rules: "everyone does it," "the system's too strict," "we had to keep the schedule." Once that justification settles, discomfort fades. Confirmation bias seals it: every smooth flight without incident feels like proof the shortcut works (Festinger, 1957; Reason, 1997).

Organizational pressures add another layer. When efficiency demands never stop, safety goals quietly fade – what Hollnagel, Woods and Leveson (2006) call **goal sedation.** Crews, engineers, and managers start prioritizing schedule or output, often unconsciously, over vigilance. Without open reporting or challenge, drift spreads silently through teams.

The danger is that drift hides in plain sight. It can persist for years, until a chain of normalized deviations meets the wrong conditions and triggers failure. Over time, safety alignment can quietly erode through shortcuts and unchecked assumptions, or strengthen through reflection, openness, and feedback. These two contrasting paths, one toward complacency, the other toward resilience, are shown conceptually in Figure 2.1, illustrating how cultural drift and learning-oriented adaptation evolve over time.

Culture Isn't Built in Crises

It's often said that crises reveal true character. In aviation, they more accurately reveal cultural preparation. A crew or company's ability to uphold safety under pressure doesn't come from adrenaline – it comes from months or years of habits, trust, and daily reinforcement (Zohar, 1980).

Research into organizational resilience shows that high-performing teams demonstrate *latent readiness* – a state where shared mental models, mutual trust,

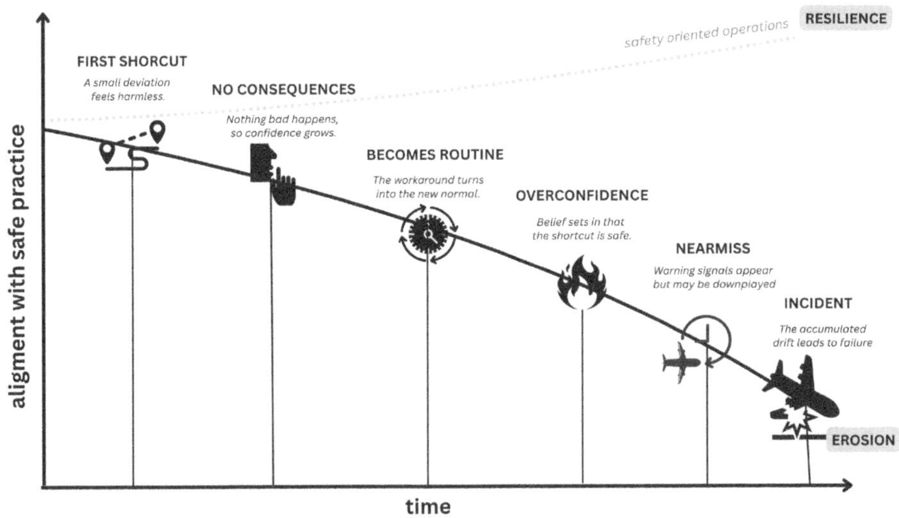

FIGURE 2.1 Safety culture drift over time (normalization of deviance).

and fluid coordination are already in place before a crisis ever occurs (Weick and Sutcliffe, 2007). These qualities don't suddenly appear in the simulator. They grow out of everyday operations: informal mentoring, how feedback is given, how leaders react to small errors, and whether reporting feels safe.

From a psychological standpoint, this aligns with the **habituation of safety behaviors** – the way repeated proactive actions become second nature (Neal and Griffin, 2006). It also connects with the idea of **psychological safety** (Edmondson, 1999): the confidence that one can speak up, challenge, or admit mistakes without fear of blame or exclusion.

When these elements are present, culture cushions teams in crises. It gives them bandwidth to adapt rather than freeze, to coordinate rather than fragment. But when culture is brittle, crises only expose the hollow slogans. A safety mindset cannot be improvised in the moment – it has to be lived every day.

WHEN SAFETY CULTURE IS TESTED: THE ROLE OF OPERATIONAL PRESSURE

Safety culture feels simple when skies are clear: when there's time to brief, when everyone is rested, when the plan holds. But that's not when it's truly tested. The real measure comes when the clock is ticking, fatigue meets turbulence, or the pressure to deliver quietly outweighs the instinct to pause.

In aviation, operational pressure isn't rare; it's built into the system: weather delays, roster gaps, maintenance backlogs, passenger impatience. Pressure doesn't just test individuals; it tests the beliefs that hold an organization together (Dekker, 2012; Patankar et al., 2012). Under that weight, culture seldom collapses dramatically, it bends in smaller ways. Focus narrows. Trade-offs fade from view. A First Officer holds back a concern to keep on schedule. A mechanic signs off "just this once." A supervisor avoids the hard conversation because disruption feels riskier than honesty.

Pressure doesn't always cause failure, but it exposes what kind of safety culture truly exists beneath the procedures.

The Erosion Effect

The erosion of safety culture rarely announces itself – it arrives disguised as practicality. A technician learns which steps can be skipped "when it's busy." A scheduler quietly penalizes a pilot for calling fatigued. Each small signal teaches people what's truly valued, and soon the spoken culture and the lived culture drift apart.

It's more than habit; it's gradual cognitive recalibration. When people are repeatedly exposed to risky behavior, they become less sensitive to it. This is known as **cognitive erosion** (Baumeister et al., 2007). Over time, the brain normalizes what it sees most. Fatigue deepens the effect: sleep-deprived individuals show reduced activity in the prefrontal cortex, the area governing judgment and ethical reasoning (Goel et al., 2009; Kusev et al., 2017).

Culture also strengthens what it tolerates. When respected figures model shortcuts or dismiss safety barriers, those around them absorb it as acceptable. Soon, the exceptions become the rules. This quiet shift spreads through maintenance teams, flight decks, or scheduling offices – anywhere questioning authority feels risky or pointless (Zohar, 2002).

Culture doesn't usually erode in crises; it erodes on ordinary days, under ordinary stress, while everyone is just trying to keep up.

Leadership under Load

True leadership isn't measured when everything runs smoothly – it's defined when the margin for error disappears (Zohar, 2002). Leaders who protect the "safety space" during peak pressure preserve not only performance but trust. Those who compress that space to chase targets –intentionally or not – send a clear message about what really matters.

Leadership under load is less about speeches or calm briefings and more about micro-behaviors: how a line captain responds when someone raises a concern, how a chief pilot frames an operational delay, or how a manager reacts when a fatigue call disrupts the schedule. Every response signals the boundaries of safety permission.

Effective leaders don't just keep people informed – they buffer them. They absorb organizational pressure, regulate their own stress responses, and create psychological room for their teams to think clearly. **Emotional intelligence** and **self-regulation,** both linked to resilience under pressure, are central to this (Goleman, 1995; Salovey and Mayer, 1990).

Conversely, when leaders withdraw, micromanage, or assign blame under stress, culture deteriorates quickly. Fear replaces openness, compliance replaces communication, and people start performing safety rather than practicing it.

Leadership under load is about presence, listening, and courage – the courage to say: This can wait. Safety can't.

CULTURAL INDICATORS: HOW TO SPOT A HEALTHY OR FAILING SAFETY CULTURE

You can't check the fuel gauge for safety culture. There's no cockpit dial or dashboard light in the crew room that flashes green when the culture's strong, or red when it's eroding. But that doesn't mean you can't sense it. Culture is everywhere – woven into pauses before someone speaks, the way eyes meet, how a doubt is phrased, or whether it's voiced at all.

It lives in the tone of a pre-flight briefing, the rhythm of a debrief, and in who gets invited to speak – and who doesn't. It shows itself in how leaders respond to discomfort and how team members handle uncertainty together. Most of all, it breathes in the ordinary moments between procedures – when the SOP runs out of script and people must rely on trust.

In healthy environments, culture doesn't have to be loud or polished. It's felt in the quiet confidence of a team that can ask hard questions without fear. Where people lean in, not away, when something feels off. Where challenge isn't confrontation – it's care.

In less healthy places, everything can look fine on the surface. Nobody argues. Nothing explodes. But there's a stiffness in the air – a hesitation. People know the rules but don't speak when the rules no longer fit the reality. The unspoken rule becomes: *don't cause waves.* And so, risk doesn't roar – it simmers (Dekker, 2019; ICAO, 2018; Martyka and Lebecki, 2014; Reason, 1997).

Reading the Signs of Culture

So, how do we actually read the room? Not with dashboards. With ears and eyes.

You'll see it in the engineer who flags a near miss without being prompted. The First Officer who says, "Are we sure we want to go ahead with this?" even if it means delaying. The cabin crew member who gently corrects a colleague, regardless of stripes. And the Captain, who, before pushback, says: "If I miss something, say it."

These small moments are not trivial – they're data. And they're everywhere if you're looking.

Signs of a Thriving Culture

- Engineers and ground crew volunteering feedback – without fear of blowback.
- Junior staff offering input and being heard, not just tolerated.
- Pilots openly discussing errors in post-flight debriefs without defensiveness.
- Supervisors asking, not assuming: "What do you need to do this safely?"

Signs of a Brittle Culture

- Hesitation during briefings; silence during debriefings.
- Politeness masking uncertainty – nobody wants to look "difficult."

- Reports decreasing, but not because the risk has improved – because trust has eroded.
- Phrases like "This is how we do it here" being used to shut down ideas, not open up dialogue.

Culture is not what you say on the safety banner. It's what happens at 04:00 on a short turnaround when the checklist doesn't quite match reality.

Listening beyond Words

What matters most in culture is often what's not said, like the sigh from the back of the room, the glance before someone answers, the way a person softens their tone before speaking in a safety meeting; not because they doubt the facts, but because they're unsure it's safe to say them. The awkward silence after someone asks, "Should we really be doing it this way?" isn't empty; it's the pulse of the organization (Akselsson et al., 2009).

Yet many operators still track spreadsheets more closely than they track silences. Metrics can mislead: a drop in fatigue reports doesn't always mean crews are sleeping better; it may mean they've stopped believing anyone's listening.

True leadership begins not with targets but with curiosity. It lives in the questions: "Why didn't we hear about that?" "Why do the same few voices speak in every meeting?" "Why did everyone go quiet when she mentioned rostering?"

If we want a genuine safety culture, we have to be fluent in more than policy. We must listen to discomfort, notice avoidance, and read not just the words but the tone (Dekker, 2012). Culture isn't declared – it's detected, one conversation, one hesitation, one brave question at a time.

Assessing for Future Safety: The Human Component

When people talk about safety culture, they often focus on systems, procedures, or leadership. But there's another layer – one that sits quietly within the individual. No matter how robust an organization's framework is, if the people inside don't personally value safety, something will eventually slip through the cracks.

Research consistently shows that attitudes toward safety – especially risk perception, sense of responsibility, and communication openness – are shaped by stable **personality traits** (Mearns et al., 2003; Neal and Griffin, 2006). Training can improve behavior, but deep-seated **attitudes** and **coping styles** are harder to shift, especially under stress (Harvey et al., 2002). You can teach checklists, but you can't easily teach *care.*

That's why assessing **individual safety attitudes** during selection and promotion is gaining traction across high-reliability industries (Su, 2021). Psychometric profiling, values-based interviews, and behaviorally anchored simulations reveal whether someone is likely to speak up, anticipate risk, or prioritize collective safety over speed (Nieva and Sorra, 2003). These aren't "soft skills." They're safety assets.

If we start looking not only at systems but at the people within them, we can build teams where safety isn't just enforced – it's *believed in.* A strong safety culture grows from both ends: leadership and systems from the top, and values and mindset from the individuals who live it daily.

To future-proof safety, we must move beyond compliance and start asking the deeper question: *Does this person believe in safety when no one's watching?*

BUILDING A SAFETY MINDSET: PRACTICAL TOOLS AND EVERYDAY BEHAVIORS

A genuine safety mindset isn't abstract but practiced daily. It lives in how people speak up, how leaders respond, and how feedback turns into learning. It thrives where openness replaces fear and where doing the right thing feels routine, not heroic.

Leadership Modeling and Psychological Safety

The foundation of any safety mindset is **psychological safety**. If a First Officer can't say, "That heading doesn't look right," or an engineer fears being labeled difficult for flagging a concern, no checklist will save the operation. As Edmondson (1999) showed, psychological safety enables honesty and learning – especially when stakes are high.

Leadership modeling anchors this principle (Bastola, 2020; Simard and Marchand, 1994). When a senior captain admits an error or a maintenance supervisor encourages questions, it signals that vulnerability is professional, not weak. Such a behavior can be deliberately cultivated – through command upgrades, instructor training, and recurrent courses (Bastola, 2020). Even brief examples in manuals or line briefings can help normalize open communication and constructive challenge.

Feedback Reinforcement

Mindset without structure remains philosophy. Feedback loops must invite reflection, not fear (Flin, 1998). Strong organizations ask not only *what went wrong* but *what nearly went wrong* – and also*: why it didn't.* Feedback should be regular, specific, and taught as a skill rather than left to personality or chance.

Incentives also quietly shape what people value. When efficiency is rewarded over judgment, compliance replaces curiosity. Simple changes, like adding a "safety mindset" comment box in post-flight reports or peer recognition for proactive behavior, reinforce the right habits.

The Southwest Flight 1380 emergency made this visible. The crew's calm and composure weren't luck; they reflected habit, trust, and a culture where safety wasn't a checklist item but a mindset lived every day.

Integrity in Practice

No safety culture is tested in calm air. Its strength reveals itself when time is short, resources are thin, and demands collide. Even the best systems fray under pressure.

What holds them together is the intersection of leadership behavior, organizational support, and team resilience.

Safety culture integrity shifts with every decision made under load. When leaders remain present and connected, it strengthens; when pressure silences communication, it weakens.

This balance is illustrated in Figure 2.2, which depicts the dynamic relationship between leadership, operational pressure, and resilience. It shows how safety culture holds when leaders buffer their teams and communication flows, and how it fractures when overload or silence takes its place.

Culture doesn't shift overnight, but each micro-moment matters. Every open conversation, every corrected shortcut, and every lesson honestly shared builds the kind of mindset that doesn't just follow safety rules but lives them. A safety culture isn't declared. It's detected – over time, through patterns, silences, and subtle acts of trust. And while systems, technology and procedures can guide, true safety begins with people: their personalities, values, and attitudes. The more individuals we have who naturally prioritize safety, the stronger the foundation becomes. But even with the right traits, culture must still be nurtured – taught, modelled, reinforced, and supported by everyday tools, shared language, and consistent leadership from the top (this mirrors as well the layered interaction between internal and external factors introduced in Chapter 1 (SPINAL-CORD model).

FIGURE 2.2 Interplay between leadership, operational pressure, and organizational resilience in shaping safety culture integrity.

TAKEAWAYS

For Pilots and Cabin Crew

- Speak up early – even about small concerns. Culture builds in the quiet moments, not just the big ones.
- Treat briefings and debriefings as shared spaces for trust-building, not one-way communication.
- Take a minute after each flight to reflect on whether operational pressure shaped any of your decisions.

Reflective Question: *How do I help create a cockpit or cabin environment where people feel safe to speak up-even when it's uncomfortable?*

For Instructors, Examiners, and Leaders

- Share your own safety lessons, including times when you missed something and what you learned.
- Reinforce behavioral markers of safety culture during feedback and CRM discussions.
- Break silos by inviting open dialogue between technical roles (engineers, cabin crew, pilots) in training environments.

Reflective Question: *Do I model safety culture in what I do, or only in what I say? What can I do to show it more?*

For Operational Teams

- Make safety culture visible – through dashboards, stories, post-incident reflections, and team rituals.
- Recognize individuals not just for avoiding errors, but for promoting trust, speaking up, and staying curious.
- Embed safety mindset principles into all phases of training, not just during CRM modules.

Reflective Question: *If reporting dropped in our organization, would we treat that as a success – or a warning sign?*

REFERENCES

Accident Investigation Bureau Nigeria (AIB Nigeria) (2017) *Accident Report on Dana Airlines Limited, Boeing MD-83, 5N-RAM, Which Occurred at Iju-Ishaga, Lagos, Nigeria on 3rd June, 2012*. Abuja: Federal Ministry of Aviation. Available at: https://asn.flightsafety.org/reports/2012/20120603_MD83_5N-RAM.pdf

Aircraft Accident Investigation Commission of Nepal (AAIC) (2023) *Final Report: Accident to ATR 72-212A, 9N-ANC, Yeti Airlines Flight 691, Pokhara, Nepal, 15 January 2023*. Kathmandu: AAIC. Available at: https://bea.aero/fileadmin/user_upload/9N-ANC_FINAL_Report.pdf

Akselsson, R., Ek, Å., Koornneef, F., Stewart, S., & Ward, M. (2009). Resilience Safety Culture. In *17th World Congress on Ergonomics 2009 IEA*. Available at: https://lucris.lub.lu.se/ws/portalfiles/portal/5444743/1608661.pdf

Aven, T. (2022) 'A risk science perspective on the discussion concerning Safety I, Safety II and Safety III', *Reliability Engineering & System Safety*, 217, p. 108077. https://doi.org/10.1016/j.ress.2021.108077

Bastola, D. P. (2020) 'The relationship between leadership styles and aviation safety: A study of the aviation industry', *Journal of Air Transport Studies*, 11(1), pp. 71–96.

Baumeister, R. F., Vohs, K. D. and Tice, D. M. (2007) 'The strength model of self-control', *Current Directions in Psychological Science*, 16(6), pp. 351–355. https://doi.org/10.1111/j.1467-8721.2007.00534.x

Dekker, S. (2012) *Just Culture: Balancing Safety and Accountability*. 2nd ed. Boca Raton, FL: CRC Press. https://doi.org/10.4324/9781315251271

Dekker, S. (2014) *The Field Guide to Understanding 'Human Error'*. 3rd ed. Boca Raton, FL: CRC Press. https://doi.org/10.1201/9781317031833

Dekker, S. (2019) *Foundations of Safety Science*. Boca Raton, FL: CRC Press.

European Union Aviation Safety Agency (EASA) (2025) *Evidence-Based Training (EBT) Manual, Version 2.3*. Cologne: EASA. Available at: https://www.easa.europa.eu/en/domains/aircrew-and-medical/evidence-based-training-ebt

Edmondson, A. (1999) 'Psychological safety and learning behavior in work teams', *Administrative Science Quarterly*, 44(2), pp. 350–383. https://doi.org/10.2307/2666999

Festinger, L. (1957) *A Theory of Cognitive Dissonance*. Stanford, CA: Stanford University Press.

Flin, R. (1998) 'Safety condition monitoring. Lessons from 'man-made disasters'', *Journal of Contingencies and Crisis Management*, 6, pp. 88–92.

Goel, N., Rao, H., Durmer, J. S. and Dinges, D. F. (2009) 'Neurocognitive consequences of sleep deprivation', *Seminars in Neurology*, 29(4), pp. 320–339. https://doi.org/10.1055/s-0029-1237117

Goleman, D. (1995) *Emotional Intelligence: Why It Can Matter More than IQ*. New York: Bantam Books.

Harvey, J., Erdos, G., Bolam, H. and Cox, M. (2002) 'An analysis of safety culture attitudes in a highly regulated environment', *Work & Stress*, 16(1), pp. 18–36. https://doi.org/10.1080/02678370110113226

Helmreich, R. L. and Merritt, A. C. (2001). *Culture at Work in Aviation and Medicine: National, Organizational and Professional Influences*. 1st ed. Abingdon: Routledge. https://doi.org/10.4324/9781315258690

Hollnagel, E. (2009) 'The four cornerstones of resilience engineering', in C. P. Nemeth, E. Hollnagel, and S. Dekker (Eds.), *Resilience Engineering Perspectives: Preparation and Restoration* (pp. 117–137). Farnham: Ashgate.

Hollnagel, E. (2014) *Safety-I and Safety-II: The Past and Future of Safety Management*. Farnham: Ashgate. https://doi.org/10.1201/9781315607511

Hollnagel, E., Woods, D. D. and Leveson, N. (Eds.) (2006) *Resilience Engineering: Concepts and Precepts*. Boca Raton, FL: CRC Press. https://doi.org/10.1201/9781315605685

International Civil Aviation Organization (ICAO) (2018) *Safety Management Manual (SMM). 4rd ed. Doc 9859. Montreal.* Available at: https://store.icao.int/en/safety-management-manual-doc-9859/ https://elibrary.icao.int/product/229751

Kahneman, D. (2011) *Thinking, Fast and Slow*. New York: Farrar, Straus and Giroux.

Kusev, P., Purser, H., Heilman, R., Cooke, A. J., Van Schaik, P., Baranova, V., Martin, R. and Ayton, P. (2017) 'Understanding risky behavior: The influence of cognitive, emotional and hormonal factors on decision-making under risk', *Frontiers in Psychology*, 8, p. Article 102. https://doi.org/10.3389/fpsyg.2017.00102

Leveson, N. (2020a) *Safety-III: A Systems Approach to Safety and Resilience.* MIT Engineering Systems Division. Available at: http://sunnyday.mit.edu/safety-3.pdf

Leveson, N. (2020b) 'Safety I–II, resilience and antifragility engineering: A debate explained through an accident occurring on a mobile elevating work platform', *International Journal of Occupational Safety and Ergonomics*, 25(1), pp. 66–75. https://doi.org/10.1080/10803548.2018.1444724

Martyka, J. and Lebecki, K. T. (2014) 'Safety culture in high-risk industries', *International Journal of Occupational Safety and Ergonomics (JOSE)*, 20(4), pp. 561–572. https://doi.org/10.1080/10803548.2014.11077076

Mearns, K., Whitaker, S. M. and Flin, R. (2003) 'Safety climate, safety management practice and safety performance in offshore environments', *Safety Science*, 41(8), pp. 641–680. https://doi.org/10.1016/S0925-7535(02)00011-5

National Transportation Safety Board (NTSB) (2019) *Uncontained Engine Failure and Subsequent Depressurization, Southwest Airlines Flight 1380, Boeing 737-7H4, N772SW, Near Philadelphia, Pennsylvania, April 17, 2018.* Accident Report NTSB/AAR-19/03. Washington, DC: NTSB. Available at: https://www.ntsb.gov/investigations/AccidentReports/Reports/AAR1903.pdf

Neal, A. and Griffin, M. A. (2006) 'A study of the lagged relationships among safety climate, safety motivation, safety behavior, and accidents at the individual and group levels', *Journal of Applied Psychology*, 91(4), pp. 946–953. https://doi.org/10.1037/0021-9010.91.4.946

Nieva, V. F. and Sorra, J. (2003). 'Safety culture assessment: A tool for improving patient safety in healthcare organizations', *BMJ quality & safety*, 12(suppl 2), pp. ii17–ii23. https://doi.org/10.1136/qhc.12.suppl_2.ii17

O'Connor, P., O'Dea, A., Kennedy, Q. and BuAttrey, S. E. (2011) 'Measuring safety climate in aviation: A review and recommendations for the future', *Safety Science*, 49(2), pp. 128–138. https://doi.org/10.1016/j.ssci.2010.10.001

Patankar, M. S., Brown, J. P., Sabin, E. J. and Bigda-Peyton, T. G. (2012) *Safety Culture: Building and Sustaining a Cultural Change in Aviation and Healthcare.* Farnham: Ashgate.

Reason, J. (1997) *Managing the Risks of Organizational Accidents.* 1st ed. Abingdon: Routledge. https://doi.org/10.4324/9781315543543

Salas, E., Shuffler, M. L., Thayer, A. L., Bedwell, W. L. and Lazzara, E. H. (2015) 'Understanding and improving teamwork in organizations: A scientifically based practical guide', *Human Resource Management*, 54, pp. 599–622. https://doi.org/10.1002/hrm.21628

Salovey, P. and Mayer, J. D. (1990) 'Emotional intelligence', *Imagination, Cognition and Personality*, 9(3), pp. 185–211. https://doi.org/10.2190/DUGG-P24E-52WK-6CDG

Shults, T. J. (2019). 'Nerves of steel: how I followed my dreams, earned my wings, and faced my greatest challenge'. W Publishing Group, an imprint of Thomas Nelson.

Simard, M. and Marchand, A. (1994) 'The behavior of first-line supervisors in accident prevention and effectiveness in occupational safety', *Safety Science*, 17(2), pp. 169–185. https://doi.org/10.1016/0925-7535(94)90010-8

Su, W.-J. (2021). 'The effects of safety management systems, attitude and commitment on safety behaviors and performance', *International Journal for Applied Information Management*, 1(4). https://doi.org/10.47738/ijaim.v1i4.20

UK Civil Aviation Authority (CAA) (2002) *CAP 719: Fundamental Human Factors Concepts in Civil Aviation.* London: Civil Aviation Authority. Available at: https://publicapps.caa.co.uk/modalapplication.aspx?appid=11&mode=detail&id=226

Vaughan, D. (1996) *The Challenger Launch Decision: Risky Technology, Culture, and Deviance at NASA.* Chicago, IL: University of Chicago Press.

Weick, K. E. and Sutcliffe, K. M. (2007). *Managing the Unexpected: Resilient Performance in the Age of Uncertainty.* 2nd ed. San Francisco, CA: Jossey-Bass/Wiley.

Wiegmann, D. A., Zhang, H., Von Thaden, T., Sharma, G. and Mitchell, S. (2002). A synthesis of safety culture and safety climate research (Technical Report No. ARL-02-3/FAA-02-2). Federal Aviation Administration. https://www.nrc.gov/docs/ML1025/ML102500649.pdf

Zohar, D. (1980) 'Safety climate in industrial organizations: Theoretical and applied applications', *Journal of Applied Psychology*, 65(1), pp. 96–102. https://doi.org/10.1037/0021-9010.65.1.96

Zohar, D. (2002) 'Modifying supervisory practices to improve subunit safety: A leadership-based intervention model', *Journal of Applied Psychology*, 87(1), pp. 156–163. https://doi.org/10.1037/0021-9010.87.1.156

Zohar, D. (2010) 'Thirty years of safety climate research: Reflections and future directions', *Accident; analysis and prevention*, 42(5), pp. 1517–1522. https://doi.org/10.1016/j.aap.2009.12.019

3 Mental Health and Well-Being in Aviation

A Hidden Risk, A Shared Responsibility

CASE STUDY: GERMANWINGS FLIGHT 9525 – THE COST OF SILENCE (2015)

It was a clear spring morning over the French Alps on March 24, 2015 when Germanwings Flight 9525 (D-AIPX) lifted off from Barcelona-El Prat Airport, bound for Düsseldorf. The Airbus A320-211, operated by Germanwings (a Lufthansa subsidiary), climbed through calm skies with 144 passengers and six crew members on board (BEA, 2016).

At about 09:30 local time, while cruising at FL380, the captain left the flight deck for a short break – a routine procedure performed countless times each day. Within moments, the sequence turned irreversible. First Officer Andreas Lubitz manually selected a descent altitude of 100 ft and locked the cockpit door from the inside. The autopilot initiated descent mode.

Air traffic controllers attempted contact after noticing the unexpected descent. Inside, the cockpit voice recorder (CVR) captured the captain knocking, then pounding on the reinforced door, along with audible alarms and steady breathing from Lubitz. There was no response. Approximately 10 minutes later, at 10:41 local time, the aircraft impacted terrain near Prads-Haute-Bléone, in the French Alps, killing all 150 people on board (BEA, 2016).

The Bureau d'Enquêtes et d'Analyses (BEA) investigation concluded that the first officer intentionally initiated the descent and that the accident resulted from deliberate action (BEA, 2016).

Beyond the immediate sequence of events, the investigation pointed to a wider context characterized by fragmented information and the absence of effective mechanisms for intervention. Indicators of serious psychological distress (including a documented history of depressive episodes, repeated medical consultations for non-specific symptoms, periods of declared unfitness to fly, and suspected psychotic depression) had been visible across time and across multiple professional interfaces, yet remained disconnected or "invisible" for the system. Medical confidentiality requirements, procedural boundaries, and the lack of clear escalation pathways meant that concerns were never integrated into a shared assessment of risk. Viewed retrospectively, these elements

DOI: 10.1201/9781003515418-3

form a coherent narrative; viewed prospectively, they appeared isolated and insufficient to trigger decisive action. The accident exposed not a single point of failure, but the consequences of silence within systems designed to protect privacy rather than manage cumulative risk (BEA, 2016).

INTRODUCTION

Ask a civil pilot what "fit to fly" means and you'll likely hear about eyesight, blood pressure, or electrocardiograms– numbers on a page. For years, regulators believed that was enough. If psychology appeared at all, it was a checkbox during training or the annual medical. Then came March 24, 2015. Germanwings Flight 9525 shattered that illusion, forcing aviation to face an unspoken truth: untreated psychological distress can endanger safety as surely as a failed pump or cracked spar (BEA, 2016).

Investigators found that First Officer Andreas Lubitz had struggled with depression since 2008, beginning during Lufthansa training. Removed, treated, and cleared to return, he concealed parts of his medical history. Competent on the line, his condition quietly worsened. He visited several doctors for vague symptoms; some declared him unfit for duty, but he hid their notes. Toxicology revealed antidepressants and anxiolytics. One clinician suspected psychotic depression and advised hospitalization, which he refused (BEA, 2016). The fragments were visible to many, yet no one connected them.

At the time, German privacy law sealed mental health data unless a patient posed an imminent threat. Lubitz never met that threshold. The system, working exactly as written, stayed silent (BEA, 2016; EASA, 2016a).

The shock went beyond the tragedy itself. It exposed a known history, visible decline, multiple touchpoints, and no trigger for action. If that can happen, what does "fit to fly" really mean and who does it protect?

The tragedy triggered one of the most significant policy shifts in modern aviation safety. Early in 2016, European Union Aviation Safety Agency (EASA) issued **ED Decision 2016/008/R**, introducing practical guidance for airlines on how to assess pilots psychologically and provide structured peer support (EASA, 2016a). Later that year, the **Action Plan for Pilot Mental Fitness** took those ideas further, turning them into a wider commitment to well-being across the industry (EASA, 2016b). Two years on, the vision became law through **Commission Regulation (EU) 2018/1042**, which made psychological assessment, peer-support programs (PSPs), and alcohol testing mandatory for all European commercial operators (EASA, 2018).

Airlines revised cockpit access and mental health policies. Progress, yes – but policy alone doesn't erase fear. Pilots still hesitate, not from denial but from knowing what disclosure can bring: grounding, investigation, uncertainty. The Federal Aviation Administration's (FAA's) 2024 Mental Health Aviation Rulemaking Committee (ARC) called it clearly – fear of disclosure is a safety risk, and a non-punitive path is essential (FAA, 2024).

And it's not only pilots. Cabin crew can face burnout from rosters and emotional labor; controllers manage relentless cognitive load; engineers describe stigma as steady as torque checks (Flight Safety Foundation, 2024; Fredes-Collarte et al., 2024).

Because mental health in aviation isn't separate from the system – it *is* the system.

THE MENTAL HEALTH CONTINUUM – HOW MIND, BODY, AND BIOLOGY INTERACT IN AVIATION

Mental health in aviation cannot be reduced to checkboxes or certificates. It is a moving spectrum shaped by workload, sleep, environment, and connection – constantly influenced by the body's own biology. What we call "mental health" is not separate from physiology; it *is* physiology in action. When rest, rhythm, and emotional regulation align, performance stays stable. When they drift, even minor stressors can distort perception and judgment.

Understanding this continuum, from everyday well-being, through temporary distress, to clinical illness, allows aviation to move from fear and stigma to awareness and management. It means viewing psychological health not as a static label but as a dynamic state that shifts with biology, workload, and support (ICAO, 2022; WHO, 2022).

Well-Being – The Foundation Layer

Well-being forms the stable baseline beneath performance: sleep that restores, mood that resets, and energy that lasts through a duty block. A pilot can feel anxious before a check ride and still be safe to fly; a cabin crew member can finish a long-haul drained and recover with rest. Aviation systems often miss this nuance, treating health as binary: "fit" or "unfit." Yet, well-being is dynamic, influenced by fatigue, workload, circadian rhythm, and social context – all manageable if recognized early (Flight Safety Foundation, 2024; ICAO, 2022).

From a neurobiological point of view, stability depends on rhythm and recovery. Consistent sleep and daylight exposure strengthen **prefrontal regulation** – the brain's command center for reasoning and composure (Yoo et al., 2007). Physical activity and short movement bursts increase **brain-derived neurotrophic factor (BDNF)**, supporting learning and emotional balance (Erickson et al., 2011). Even controlled breathing can rebalance the autonomic system, reflected in higher **heart-rate variability (HRV)** – a marker of resilience and flexible emotional control (Laborde et al., 2017; Thayer et al., 2012).

Temporary Distress – The Normal Human Response

Temporary distress sits at the lighter end of the continuum: short-lived reactions to identifiable stressors: a run of morning duties, tension at home, or a confidence dip after a demanding flight. These are not illnesses but natural fluctuations that typically resolve with rest, time, or a short conversation. Early acknowledgment allows proportionate action, like a lighter roster, rest day, or brief peer check – rather than escalation (Casey et al., 2025; De Kock et al., 2021).

From a neurobiological perspective, short-term stress can actually enhance alertness. The body's stress system releases **cortisol** and **adrenaline** to sharpen attention and coordination (Arnsten, 2009). The key lies in recovery: when sleep and social connection return, stress hormones stabilize, cognitive clarity rebounds, and decision-making improves (Lim and Dinges, 2010). Without that reset, the same activation that once helped performance starts to blur focus and shorten patience.

Mental Illness – When Symptoms Do and Don't Mean "Unsafe"

Further along the spectrum are **diagnosable conditions** that alter how a person feels, thinks, or functions. A diagnosis, however, does not automatically mean "unfit." Operational concern arises only when symptoms (or side effects of treatment) impair concentration, judgment, or situational awareness. Many professionals manage anxiety or depression safely under medical oversight (EASA, 2016b FAA, 2024). Modern regulators distinguish between untreated illness and stable, treated conditions; with insight and regular follow-up, return to duty is often safe and appropriate.

From a biological standpoint, prolonged stress and disrupted sleep gradually reshape brain function. Sustained cortisol activation weakens prefrontal regulation, raises inflammation, and reduces neuroplasticity, setting the stage for persistent low mood or anxiety (Miller and Raison, 2016). At the same time, the brain's chemistry shifts: **dopamine**, central to motivation and reward, declines, while **serotonin** and **norepinephrine** systems lose balance. This biochemical drift compounds stress and fatigue, dulling energy and narrowing perspective (Nestler and Carlezon, 2006).

Effective treatment, rest, and consistent recovery behaviors can gradually reverse many of these effects. As sleep and activity normalize, neuroplasticity strengthens and neurotransmitter balance begins to stabilize, restoring both cognitive and emotional flexibility. Healing the brain, like retraining a skill, depends on repetition, structured feedback, and time.

AVIATION-RELATED CHALLENGES

Aviation professionals operate within environments that reward precision, speed, and control. Yet beneath that surface lies a steady current of psychological demand. Long hours, time pressure, and continuous evaluation are woven into daily life, often in ways that quietly shape attention, mood, and relationships. Understanding these pressures is essential not only for flight safety but also for sustaining a healthy, motivated workforce.

Occupational Stressors Unique to Aviation

Aviation can appear effortless from the outside: polished crews, calm voices, predictable routines, but behind that façade is a complex system where human performance is constantly stretched. Every role carries its own stress pattern: pilots juggle workload peaks and circadian disruption; cabin crews manage emotional labor in confined spaces; controllers sustain concentration under unbroken vigilance; engineers and dispatchers solve problems under noise, weather, and time pressure. (Cahill et al., 2020; Fredes-Collarte et al., 2024; Goode, 2003).

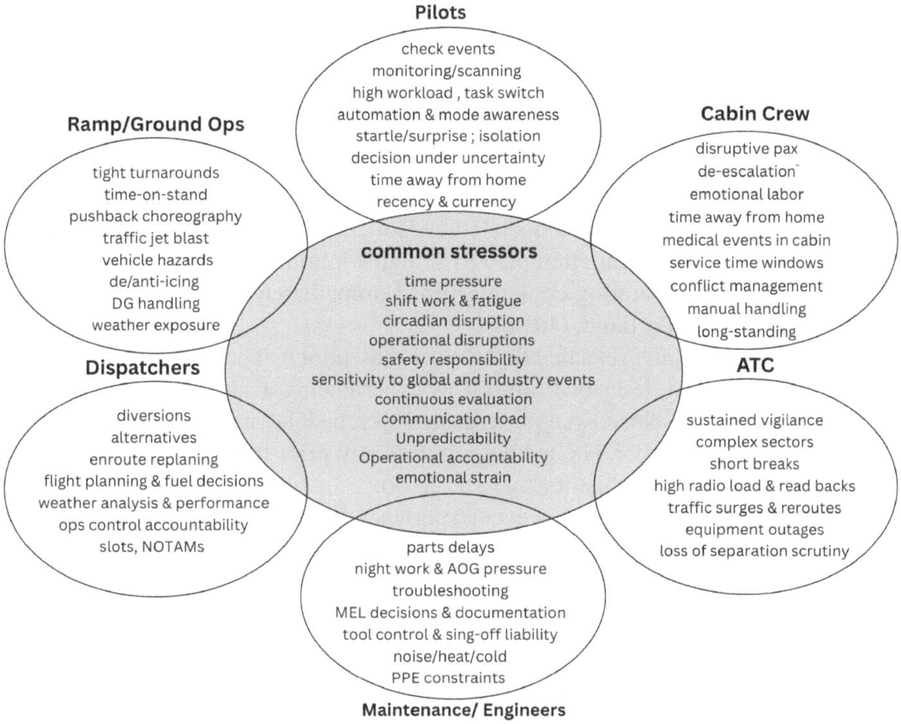

FIGURE 3.1 Occupational stressors across aviation roles.

As shown in Figure 3.1, these overlapping stressors link flight decks, cabins, towers, and hangars into one interdependent human system. Being aware of them strengthens not only operational understanding but also shared responsibility for well-being and care across roles.

Culture adds a subtler layer. Aviation still prizes composure, and for good reason, but that same ideal can turn openness into perceived weakness. Saying "I'm not sleeping well" or "I'm finding it hard to switch off" can still sound risky. That is how routine pressure becomes invisible until it escalates. A strong safety culture guards not just against operational failure but against silence. When leaders frame well-being as professionalism, not vulnerability, they remove stigma and make help-seeking routine.

The "Invincibility" Myth and the Legacy of Stigma

Resilience is one of aviation's greatest strengths. Silence isn't. For decades, the image of the perfect pilot was fixed – calm voice, steady hands, emotions parked at the door. The culture rewarded composure, not conversation, and training reinforced it.

The result is what people don't say. Pilots learn to "push through," to manage privately, to wait for it to pass. Yet the data tell another story. In a large international survey of 1,837 airline pilots, **12.6 percent met the criteria for depression and 4.1 percent reported suicidal thoughts in the previous two weeks**, with very few ever disclosing their symptoms to employers or regulators (Wu et al., 2016). Awareness isn't the issue – the consequence is.

A recent case showed how far silence can go. On October 22, 2023, during **Alaska Airlines Flight 2059** operated by Horizon Air, an off-duty pilot riding the flight deck jump seat attempted to shut down both engines by pulling the fire handles. The operating crew intervened immediately, restrained him, and diverted safely to Portland, Oregon.

Investigations later revealed that the off-duty pilot had been awake for more than 40 hours and described himself as experiencing a severe mental health crisis. He also disclosed having ingested psychedelic mushrooms approximately two days earlier. He had not reported any prior mental health difficulties in his aeromedical history (NTSB, 2024).

It was a harsh reminder that even experienced professionals can deteriorate quietly when fear of judgment outweighs trust in the system.

EASA's post-Germanwings guidance was built to prevent exactly this – access to help without automatic career loss (EASA, 2016b). Yet, stigma endures. **Emotional suppression** – so keeping the cockpit tone flat while stress runs high – raises physiological strain and burnout risk (Hülsheger and Schewe, 2011). Across safety-critical jobs, stigma still delays care (Henderson et al., 2013). The "invincibility" myth doesn't protect performance; it quietly erodes it.

The "Gen Z" Era and the Post-Pandemic Shift

A new generation is now entering cockpits and cabins with different expectations. They value psychological openness and balance yet face a volatile industry shaped by global disruption. From an industry standpoint, the pandemic left deep marks: training pipelines froze, cadets waited months for type-rating calls, and newly qualified pilots struggled to retain proficiency (Olaganathan and Amihan, 2021). The pause blurred professional identity just as many were forming it.

Generation Z professionals speak more freely about mental health but also report higher baseline anxiety and sleep disturbance linked to constant digital engagement (Twenge et al., 2019). They expect clarity – who to contact, what happens after disclosure, and whether support will be fair. For airlines and training organizations, this generation's openness is an opportunity to modernize culture (Bender et al., 2025)

The post-pandemic workforce also forced a rethink of resilience training. Traditional Crew Resource Management (CRM) language around "toughness" now feels outdated. Instead, line checks, briefings, and recurrent training increasingly include discussions about attention, fatigue, and coping as integral to performance.

When organizations meet transparency with structure, rather than slogans, they gain a generation that sees self-care as part of safety – not apart from it.

SYSTEMS, SAFEGUARDS, AND GAPS

Aviation safety systems are built on the idea that every risk can be managed through structure. But when it comes to mental health, regulation, and culture don't always move in sync. After Germanwings, the industry made substantial progress – introducing psychological assessments, two-person cockpit policies, and PSPs, but implementation across regions has been uneven. The challenge now is not awareness, but consistency: how to ensure that every operator, regardless of geography or size, can access the same standards of psychological safety.

Global Regulatory Frameworks

As mentioned, regulatory approaches to mental health have evolved rapidly but remain fragmented. Europe, under EASA, leads with mandatory psychological assessments and PSPs. Other authorities, including the FAA and Transport Canada, have chosen guidance and voluntary participation over mandates. The result is a global mosaic: progress in many areas, gaps in others, and significant opportunity for shared learning.

These differences aren't failures, they reflect how each system balances privacy, law, and culture. In some regions, data protection laws limit how much personal information can be shared; in others, the barrier is resource availability or public perception. What matters is the common intent: to protect well-being without undermining trust. Aviation, as a global network, could benefit from structured knowledge exchange between regulators – learning from successes, not only from accidents (Table 3.1).

What this table shows is not a hierarchy but a timeline of evolution. Each region is building awareness at its own pace, shaped by law, culture, and operational need. We can only hope that, as understanding grows, regulators around the world will work toward a shared model – one that safeguards confidentiality while encouraging early, proactive support. Such alignment would make psychological safety as tangible and measurable as technical safety itself.

Peer-Support Programs: Progress, Pitfalls, and Untapped Potential

PSPs are among the most meaningful changes to follow Germanwings. They give pilots and crew a confidential space to talk with trained colleagues who understand operational realities. In those conversations, distress meets empathy instead of judgment, and help-seeking becomes a sign of responsibility, not weakness (EASA, 2023).

The strength of any PSP rests on three pillars: **trust, competence, and integration**. Trust must be absolute – if participants doubt confidentiality, the system collapses. Peer supporters need solid training in listening, boundary-setting, and referral, along with supervision to protect their own well-being. Integration means

TABLE 3.1
Global Mental-Health Oversight for Flight Crew: Status as of 2025

Region/Regulator	Status	Summary of Current Approach
FAA (the United States)	**Developing**	No mandated psychological testing; the FAA's *Mental Health Aviation Rulemaking Committee* recommends non-punitive disclosure pathways, enhanced AME training, and voluntary peer-support mechanisms, without introducing binding regulatory requirements (FAA, 2024)
Transport Canada (Canada)	**Partial**	No formal psychological testing; Transport Canada (TC) Aviation Safety Letter articles discuss mental health and psychological safety, emphasising education, CRM, and organizational awareness without regulatory mandates (TC, 2024; TC, 2025).
CASA/CAA NZ (Australia and New Zealand)	**Partial**	CASA addresses mental health through human factors and pilot performance guidance rather than SMS mandates (CASA, 2021), while CAA New Zealand's *General Directions – Temporary Medical Conditions* establish Safe Haven/TMC pathways with defined non-reporting rules, without mandatory psychological assessment (CAA NZ, 2025).
EASA (Europe, the UK, NO, CH)	**Mandatory/ developing**	European framework mandates psychological assessment at (re) employment, peer-support programmes, and substance-misuse policies for all commercial air operators, introduced following Germanwings and embedded in EU law through Commission Regulation (EU) 2018/1042, with ongoing evaluation and guidance by EASA (EASA, 2016a, 2016b, 2023; EU, 2018)
DGCA (India)	**Mandatory/ developing**	*Medical Circular 01 of 2023* mandates psychological assessment, peer support, and mental-health promotion within SMS; implementation underway across operators (DGCA, 2023)
APAC Region (Asia-Pacific)	**Developing. mixed adoption**	Operator-led programs in Japan (Japan Airlines [JAL]), All Nippon Airways [ANA] and Singapore's Tripartite Framework for pilots and air traffic control officers introduce structured peer support and education; broader ICAO-region uptake – including China – remains voluntary and state-driven (ANA Group, 2023; CAAS, 2023; ICAO, 2023).
LATAM (South America and Caribbean)	**Developing/ mixed adoption**	Mental-health and wellbeing initiatives are emerging primarily through airline-led peer-support programmes and regional safety initiatives, with regulatory integration still evolving (ANAC, 2023; ICAO SAM, 2023)
Africa and Middle East	**Emerging/ partial**	Mental-health considerations mainly addressed through aeromedical fitness and human-factors frameworks, supplemented by voluntary peer-support initiatives in Africa (e.g. guidance from the South African Civil Aviation Authority and support provided by Mayday-SA); regulator-led education and substance-misuse programmes in the United Arab Emirates under the General Civil Aviation Authority, with limited formal, region-wide wellbeing regulation (GCAA UAE, 2024; Mayday, 2025; SACAA, 2023)

linking the program with Safety Management Systems (SMS) and occupational health pathways, so issues can be addressed early without breaching confidentiality (IATA, 2024).

When these elements align, PSPs work. Research shows they increase help-seeking, reduce crisis escalation, and foster a sense of belonging across teams (Melin and Lång, 2023). However, many programs still serve pilots only. Expanding them to include cabin crew, controllers, and maintenance staff would reflect how distress actually appears in operations – across the entire system, not one role.

Ultimately, PSPs remind aviation of a simple truth: connection is a safety defense. When people have a trusted peer to call before fatigue, grief, or anxiety reach the line, prevention replaces reaction.

Organizational Responsibility: From Paper to Practice

Policies alone don't build resilience; daily behavior does. Some airlines now go beyond compliance – introducing well-being action plans, resilience workshops, and structured return-to-duty processes. Others have created confidential helplines or integrated mental fitness modules into CRM. These steps matter, but consistency remains uneven, and many initiatives fade when resources tighten or leadership changes.

The most effective organizations treat mental health as a **core safety element**, not an Human Resources formality. They weave it into scheduling, rostering, and feedback processes – where fatigue, stress, and life circumstances are discussed as operational factors, not personal flaws. Support becomes visible when check-ins happen after significant events, when open discussions follow incidents, and when the question "How are you doing?" carries real intent, not courtesy.

Leadership tone is pivotal. When captains and managers speak about their own coping strategies or past challenges, they normalize the conversation and build trust. People listen differently when openness comes from those they respect. A culture that values reflection and recovery, alongside precision and performance, prevents silence from turning into risk. A workplace that punishes honesty is unsafe; one that protects it is resilient – and future-ready.

Psychological Assessment: Profiling Potential, Not Mental Health

Psychological assessment in aviation is often misunderstood as a mental health screening, but its purpose is very different. While mental health evaluations aim to detect clinical symptoms or temporary distress, psychological assessments focus on stable traits, coping styles, and behavioral tendencies that influence how a person performs under pressure. They help identify factors such as emotional regulation, stress tolerance, conscientiousness, adaptability, and teamwork orientation – traits that can shape performance, particularly in high-tempo, high-stakes environments (EASA, 2016b; Boz Semerci and Volery, 2019; Martinussen and Hunter, 2017).

Understanding these underlying dispositions provides valuable insight into how an individual is likely to respond to a challenge, not whether they are mentally unwell. For example, certain personality profiles may show a higher susceptibility to stress reactivity, perfectionism, or overcontrol – patterns that can influence coping strategies but are not in themselves indicators of pathology (Carretta, 2011; Littlepage et al., 2016).

ICAO (2022) highlights that psychological profiling and well-being monitoring should be viewed as complementary layers of safety. Personality-based assessment supports proactive risk management in recruitment and training, while well-being programs address the dynamic, situational states that fluctuate with fatigue, workload, and life circumstances. When combined, they create a holistic picture – selection for potential and stability, followed by ongoing support for adaptation and recovery.

The Role of Aviation Psychologists, Specialists, and AMEs: Still Fragmented, Still Misunderstood

Aviation psychologists and well-being specialists hold a unique position in the ecosystem of flight safety. Their work extends far beyond testing or post-incident support – they translate psychological knowledge into practical strategies that sustain performance, resilience, and trust across entire organizations (Kalaga, 2023).

Yet in many airlines and training environments, their potential remains underused. Some appear only during selection or after a major event; others are absent altogether. The result is a gap between what we know about human performance and what is actually applied on the line.

Well-being specialists come from different professional paths. Some are psychologists with aviation expertise; others are human-factors professionals or even qualified pilots trained in mental health and resilience coaching. What matters is not the title but the skill set, an understanding of human behavior, system safety, and the operational context in which people work. When these disciplines collaborate, they provide a safety net that spans from recruitment to line operations (Ćosić et al., 2024).

In practical terms, aviation psychologists and well-being specialists can:

- Design and supervise PSPs, ensuring confidentiality and professional standards.
- Support **mental-fitness education** and stress-regulation training within CRM and type-rating courses.
- Contribute to **incident reviews** by interpreting the psychological dimensions of human performance, not just procedural compliance.
- Work with Aeromedical Examiners (AMEs) **and safety departments** to develop integrated monitoring frameworks for mental health, fatigue, and resilience.

Meanwhile, Aeromedical Examiners (AMEs) remain responsible for assessing fitness to fly, but their focus is often physical and time-limited. Many AMEs lack specialized psychological training and rely on brief self-reports during medical exams (CAA UK, 2023). This is not a flaw of medicine – it's a structural issue. Without collaboration between **AMEs, aviation psychologists, and well-being specialists**, subtle warning signs can slip through unnoticed.

True psychological safety requires both perspectives: medical oversight and behavioral understanding. A coordinated model, where aviation psychologists, well-being experts, and AMEs share insights while protecting confidentiality, would provide the continuous, nuanced care that modern aviation demands (Ćosić et al., 2024).

When these roles work together, they create a bridge between science and operations, between people and systems. They turn mental health from an individual concern into an organizational strength.

IMPROVING FROM THE INDIVIDUAL POINT OF VIEW

Aviation culture has long centered on technical mastery and procedural precision, but personal insight is an equally powerful safety tool. Individual awareness: of one's limits, mindset, and coping style – forms the first layer of psychological defense. No checklist or policy can replace it. Supporting mental health begins with small, practical actions that pilots, crew, and operational staff can apply daily.

Building Self-Awareness and Psychological Literacy

Self-awareness is the ability to recognize internal signals before they show up in performance. It is knowing when attention drifts, when irritability rises, or when fatigue starts to flatten reaction time. In aviation, where decision-making often depends on split-second judgment, early recognition of stress or distraction can prevent subtle errors from growing into larger ones.

Psychological literacy – the skill of understanding one's emotional and cognitive processes – remains underdeveloped in many aviation roles. Yet it is teachable. Simple reflection after duty, journaling, or structured debriefs help translate subjective feelings into observable patterns: "I make more mistakes after short rest periods," or "I lose patience quicker after high workload." Such insights transform emotion into data – information that can be acted upon before it affects performance (De Kock et al., 2021).

From a physiological standpoint, awareness also means noticing bodily cues. Elevated heart rate, shallow breathing, or muscle tension are early signals of stress activation (Thayer et al., 2012). Learning to identify these changes allows individuals to reset before pressure escalates.

Good habits sustain that awareness. Simple daily practices – **consistent sleep schedules, proper hydration, moderate caffeine, and digital boundaries before rest** – keep the brain's alertness and recovery cycles predictable. Brief end-of-day reviews or gratitude routines help shift focus from constant evaluation to reflection and closure. Over time, these small rituals reinforce stability, making awareness an effortless, automatic skill rather than a reactive response.

Self-awareness also extends outward. **Opening yourself to others** – sharing when fatigue builds, asking for feedback, or admitting when focus dips – strengthens mutual safety. Openness replaces the illusion of control with connection and trust. In aviation teams, that honesty is not vulnerability; it's professionalism.

Ultimately, self-awareness is not about self-criticism; it's about **maintenance**. Just as pilots check instruments for deviation, checking in with oneself – and with those around you – creates the feedback loop that keeps performance and well-being aligned.

PRACTICAL TOOLS FOR EVERYDAY HEALTH

Once awareness develops, tools help maintain balance. The aim isn't perfection – it's recovery. Modern technology offers discreet, supportive feedback through wearable devices that track sleep, HRV, and circadian rhythm. These data points don't replace self-reflection but complement it, helping individuals notice when their physiology and workload start to diverge.

Equally valuable are analog tools: controlled breathing, short movement breaks, and brief mindfulness exercises between flights or shifts. Research consistently shows that even two minutes of focused breathing can reset the autonomic balance, lowering cortisol and restoring clarity (Laborde et al., 2017). Micro-restoration – **pausing, stretching, hydrating** – keeps both brain and body aligned with the demands of flight operations.

One structured way to incorporate these principles is the Preflight Mental Readiness Checklist (or Well-being Preflight Checklist), verifying well-being aspects before the flight (Table 3.2). Modeled on operational checklists, it helps individuals conduct a quick, honest self-assessment before duty. The checklist doesn't diagnose; it prompts reflection. Over time, it builds routine awareness of mental, physical, and emotional readiness, just as preflight inspections build technical safety.

When individuals use tools like this consistently, they normalize self-checking as part of professional discipline. Discussing the checklist informally in crew rooms or training sessions can also spark curiosity – "What would you add?" – and create shared ownership of well-being.

Instructors and Leaders as Mental Health Role Models

Support doesn't always come from specialists. Often, it starts with someone noticing – a line trainer, a base manager, or a senior colleague who pays attention. Mentorship and leadership are practical extensions of safety culture, not separate from it.

Instructors and managers shape how trainees and teams understand pressure. Modeling composure and openness during flights, simulator sessions, or briefings helps normalize emotional awareness as part of competence (Casey et al., 2025). Sharing small lessons – what worked after a stressful duty, how to reset between sectors, or when to take a pause – can be more powerful than any formal training.

True mentorship thrives on authenticity. It's not about offering solutions but about creating space: listening without judgment, giving honest feedback, and sharing experience in a way that builds confidence rather than dependence. Good mentors don't remove challenge; they frame it. They help others interpret stress as information, not failure, and guide them toward recovery with realism and empathy (Cullen et al., 2021).

TABLE 3.2
Preflight Mental Readiness Checklist

WELL-BEING PREFLIGHT CHECKLIST

1. Cabin – mindset	**Status**
Self-assessment intent	Confirmed
Distractions (phone/notifications)	Minimized
Focus – present attention	Set
Motivation – purpose for duty	Clear
If mood unstable or intrusive thoughts persist, pause and use a breathing reset. Seek peer or manager support before sign-on.	
2. Fatigue and Sleep	**Status**
Sleep duration – last 24/48 h	Adequate
Sleep quality – not fragmented, no heavy alcohol	Acceptable
Body clock fit – duty timing vs natural alertness	Acceptable
Microsleeps/nodding – yesterday	None
3. Hydration and Fuel	**Status**
Hydration – water intake today	Sufficient
Nutrition – recent balanced meal/snack	Completed
Caffeine – within personal limits	Set
4. Physiology and Medication	**Status**
Illness symptoms (fever, pain, GI issues)	Absent
Medication – approved, no side effects	Verified
Substances – alcohol last 12 h/others	Nil
5. Headspace and Support	**Status**
Stressors – family, finance, admin	Acknowledged
Coping plan – small action for today	Defined
Connection – someone to call if needed	Identified
Stressors – family, finance, admin	Acknowledged
6. Duty Readiness	**Status**
Briefing capacity – attention and recall	Ready
Crew coordination – speak up and listen	Ready
Recovery plan – post-duty	Set
If any item is not acceptable, adequate, verified, or ready – pause, reset, and seek support before sign-on.	

Leaders who show vulnerability set a new tone. When captains, examiners, or managers speak openly about pressure, they teach that professionalism and humanity can coexist. That's how a culture of care becomes operational, not ornamental.

Psychological assessments, while important, capture only a snapshot in time. Real safety depends on what happens between those snapshots: the daily interactions that build awareness, the small gestures that restore perspective, and the trust that allows people to speak before things go wrong. Mentors, instructors, and leaders make those moments visible, turning care into a habit rather than an exception.

Creating Safe Spaces for Asking "Are You Okay?"

Peer-support systems are one layer of protection. Culture is another. The most powerful intervention isn't always a formal program – it's often a quiet, genuine "You alright?" from a colleague who listens.

A psychologically safe workplace is one where conversations about mental load, fatigue, or emotional strain are as normal as talking about fuel planning or weather. It doesn't mean turning every briefing into therapy; it means normalizing humanity in the flight deck, the cabin, and the hangar.

You don't have to be a mental health expert to help. Many people hesitate to approach someone in distress, afraid they'll say the wrong thing or make it worse. In reality, silence is far more damaging. When someone is struggling, their neurobiology and fear – sometimes reinforced by culture – can make it difficult to reach out. Approaching them with openness, discretion, and genuine care is never harmful; it often makes the difference. And if it ever feels beyond your comfort or skill level, referring to a peer supporter or professional is the right step, not a failure.

Small, deliberate actions make the difference:

- Preflight briefings that include "How are you?" alongside Meteorological Aerodrome Reports (METARs).
- Simulator debriefs that reflect on focus and headspace, not just Standard Operating Procedures (SOPs).
- Leadership messages that acknowledge personal challenges, not only successes.

In aviation, silence is often mistaken for strength. But genuine strength is the courage to speak – and the wisdom to listen.

IMPROVING FROM THE ORGANIZATIONAL POINT OF VIEW

Creating a culture that protects mental health isn't about isolated programs – it's about how organizations think, teach, and lead every day. From the first days of training to the highest levels of management, the way people talk about pressure and recovery shapes how safely they operate. The following sections look at how aviation organizations can move beyond compliance to build systems that genuinely support psychological well-being.

Beyond policies and procedures, organizations also shape mental health through what remains unspoken. Norms around fatigue, disclosure, and vulnerability are often learned informally through organizational culture and everyday interactions, which strongly influence whether wellbeing initiatives are trusted and used in practice (Cullen et al., 2021). For this reason, effective well-being programs require design and implementation by professionals with domain-specific expertise in both aviation and mental health, as well as a deep understanding of the organization itself. In an industry where stigma and fear remain prevalent, such programs cannot be expected to deliver immediate results but must be allowed to develop organically and integrate gradually into the operational culture.

EDUCATION FROM THE START: BUILDING EMOTIONAL COMPETENCE EARLY

The foundation of psychological safety is laid long before someone joins an airline. Flight schools, cabin crew academies, and maintenance training centers are where professional attitudes toward pressure, communication, and self-care begin to form. Yet most initial training still treats mental health as a side topic – something to be discussed only when problems appear.

Resilience isn't just about bouncing back after burnout; it's about staying adaptive under changing conditions. That skill can be taught. Integrating emotional regulation and self-management training into core curricula makes these abilities as natural as threat and error management. Research shows that early education in emotional performance reduces stigma and increases help-seeking across high-reliability professions (Alinejad-Tilaki et al., 2025; De Kock et al., 2021).

Simple, structured methods work best:

- **Cognitive reframing** – seeing stress as information, not danger.
- **Grounding techniques** – brief sensory resets to restore focus.
- **Micro-recovery** – short pauses between flights or simulator sessions to clear mental load.

These aren't forms of therapy; they're operational tools. Teaching them early builds professionals who can self-monitor and self-correct before stress becomes fatigue or disengagement. When students learn that mental health is part of performance, they carry that mindset into the cockpit, the tower, and the hangar.

Company Culture: From Stigma to Openness

Fear of being grounded remains one of the strongest barriers to early help-seeking in aviation. Even when awareness is high, **stigma and regulatory anxiety** continue to shape behavior (Cullen et al., 2021). A large cross-sectional study of over 5,000 pilots from the United States and Canada found that more than half reported avoiding medical care at some point due to fear of losing their flight status (Hoffman et al., 2023). The findings showed similar rates across both countries – 56% of US pilots and 55% of Canadian pilots – and confirmed that such avoidance was not driven by demographics but by perceived professional consequences.

This pattern illustrates why organizational culture matters as much as regulation. When pilots associate care-seeking with punishment rather than professionalism, stigma becomes embedded in the system itself. Shifting that culture requires leadership tone, peer visibility, and structures that make psychological openness an operational norm, not an exception.

New policies are being introduced, and this is a significant step forward, however, policies don't create culture – people do. An organization's real attitude toward well-being is revealed not in posters or memos but in daily behavior: How fatigue reports are received, how leaders respond to mistakes, and whether asking for support feels safe or risky.

A mentally healthy culture treats honesty as professionalism. Leaders at all levels should model that – acknowledging pressure, admitting limits, and showing that help-seeking is responsible, not weak. When captains or department heads speak about stress openly, it legitimizes the conversation for everyone else (Edmondson, 2018).

Embedding this mindset means weaving well-being into daily operations:

- **Pre-duty briefings** that include focus and readiness checks.
- **Post-event debriefs** that address how teams felt, not only what they did.
- **Visible leadership participation** in resilience or mental-fitness programs.

Organizations that normalize such practices report stronger learning climates and fewer safety incidents (De Kock et al., 2021; Edmondson, 2018). Simply put, psychological safety strengthens operational safety.

Programs and Practical Ideas: Beyond Compliance

Many operators now meet the basic regulatory requirements for psychological assessment and peer support. The next step is building ecosystems where mental health is monitored, supported, and developed continuously, not only assessed once a year.

Effective initiatives include:

- **Comprehensive well-being programs** designed by specialists, combining fatigue science, physical health, and mental support.
- **Routine well-being reviews** after major transitions, long rotations, or incidents.
- **Confidential digital tools** that help staff track sleep, stress, and focus-linked to in-house professionals rather than generic providers (Ćosić et al., 2024).
- **Joint workshops** co-delivered by aviation psychologists and line personnel, translating research into operational strategies (ICAO, 2022).
- **Recognition systems** that reward empathy and communication alongside technical precision.
- **Resilience programs** that strengthen capacity to adapt under pressure, recover from setbacks, and sustain performance over time. Focuses on self-awareness, emotional regulation, coping strategies, and constructive communication (EASA, 2023).

These programs work when they are credible and culture-fit – designed by people who understand aviation's pace, rosters, and human demands. Aviation psychologists and well-being experts play a vital role in converting science into practical tools that crews trust and use (Cooper and Saunders, 2024; De Kock et al., 2021).

A truly safety-driven organization plans for mental health with the same rigor it plans for maintenance and training. It monitors it, learns from it, and invests in it – because the well-being of its people is the ultimate redundancy in every safety system.

TAKEAWAYS

For Pilots and Cabin Crew

- Make self-checks as normal as preflight checks. Your mental fitness is part of your operational readiness – monitor it regularly, not just when something feels wrong.
- Notice your early warning signs – shorter patience, scattered focus, changes in sleep or appetite. These are signals, not flaws.
- Use what's available – peer support, trusted colleagues, or a short reset before duty. Small conversations early can prevent crises later.

Reflective Question: *What stories do I tell myself about strength that might keep me from reaching out when I actually need support?*

For Instructors, Examiners, and Leaders

- Model openness and balance in every session. When you name stress or fatigue out loud, you give others permission to do the same.
- Include human factors in your debriefs – how someone managed their mindset, not just how they handled procedures.
- Build quiet trust. A two-minute chat after a sim can be more protective than any checklist when it comes to mental health.

Reflective Question: What small things could I do or say to make trainees feel safe being honest about their limits?

For Operational Teams

- Embed well-being into your SMS – treat it as an operational factor, not an HR theme. Involve professionals who understand both aviation and psychology. Programs work best when they're built by people fluent in both languages.
- Teach mental health literacy early – during initial training, type ratings, and leadership programs. Normalize it from day one.

Reflective Question: *Would your current system notice a person under strain before it becomes an incident – or only after?*

REFERENCES

Agência Nacional de Aviação Civil (ANAC Brazil) (2023) *Regulamento Brasileiro da Aviação Civil (RBAC) No. 67 – Requisitos para Concessão de Certificados Médicos Aeronautas.* Brasília. Available at: https://www.anac.gov.br/assuntos/legislacao/legislacao-1/boletim-de-pessoal/2021/38s1/anexo-i-rbac-no-67-emenda-05

Alinejad-Tilaki, A., Omidvar, S., Kheirkhah, F., et al. (2025) 'Health literacy and its relationship with mental health and quality of life in freshmen students', *BMC Public Health*, 25, p. 106. https://doi.org/10.1186/s12889-024-21202-4

All Nippon Airways Co., Ltd. (ANA Group) (2023) *Health and Medical Support Initiatives.* Available at: https://www.ana.co.jp/group/en/csr/human_resources/health_care/

Arnsten, A. F. T. (2009) 'Stress signalling pathways that impair prefrontal cortex structure and function', *Nature Reviews Neuroscience*, 10(6), pp. 410–422. https://doi.org/10.1038/nrn2648

BEA (2016) *Final Report on the Accident on 24 March 2015 to the Airbus A320-211, Flight GWI18G.* Bureau d'Enquêtes et d'Analyses. Available at: https://bea.aero/uploads/tx_elydbrapports/BEA2015-0125.en-LR.pdf

Bender, M., Hayden, A., Miksa, V. and Blickensderfer, E. (2025) I*nvestigating Young Pilots' Perception, Training, and Reliance on Automation in the Flight Deck.* Proceedings of the Human Factors and Ergonomics Society Annual Meeting 69(1) https://doi.org/10.1177/10711813251360700

Boz Semerci, A. and Volery, T. (2019) 'Longitudinal investigation on personality traits and mental health relationships: The mediating role of work-family interference and enhancement', *Current Issues in Personality Psychology*, 7(3), pp. 173–188. https://doi.org/10.5114/cipp.2019.89166

CAA New Zealand (CAA NZ) (2025) *General Directions: Exceptions for Temporary Medical Conditions (GD/TMC/01/04 01). I*ssued pursuant to Clause 25(1)(b) of Schedule 2, Civil Aviation Act 2023 Effective from 29 August 2025. Available at: https://www.aviation.govt.nz/assets/licensing-and-certification/medical/general-directions-temporary-medical-conditions-.pdf?utm_source=chatgpt.com

Cahill, J., Cullen, P., Anwer, S., Gaynor, K. and Wilson, S. (2020) 'The requirements for new tools for use by pilots and the aviation industry to manage risks pertaining to work-related stress (WRS) and wellbeing', *Technologies*, 8(3), p. 40. https://doi.org/10.3390/technologies8030040

Carretta, T. R. (2011) 'Pilot candidate selection method', *Aviation Psychology and Applied Human Factors*, 1(1), pp. 3–8. https://doi.org/10.1027/2192-0923/a00002

Casey, T. W., Hu, X., Carden, C. and Lee, Q. Y. (2025) 'Sticky interventions for a sticky problem: A systematic review of recent workplace mental health stigma reduction interventions with implications for training transfer', *Journal of Safety Research*, 92, pp. 448–458. https://doi.org/10.1016/j.jsr.2025.01.003

Civil Aviation Authority of Singapore (CAAS) (2023) *Tripartite framework for pilots & ATCOs on peer support.* Available at: https://www.caas.gov.sg/docs/default-source/default-document-library/caas_tripartite-mental-health_booklet.pdf

Civil Aviation Safety Authority (CASA) (2021) *Safety behaviours: Human factors for pilots.* Canberra: CASA. Available at: https://www.casa.gov.au/sites/default/files/2021-05/safety-behaviours-human-factor-for-pilots.pdf

Cooper, R. E., Saunders, K. E., et al. (2024) 'The effectiveness, implementation, and experiences of peer support approaches for mental health: a systematic umbrella review', *BMC Medicine*, 22, p. 120. https://doi.org/10.1186/s12916-024-03260-y

Ćosić, K., Popović, S. and Wiederhold, B. K. (2024) 'Enhancing aviation safety through AI-driven mental health management for pilots and air traffic controllers', *Cyberpsychology, behavior and social networking*, 27(8), pp. 588–598. https://doi.org/10.1089/cyber.2023.0737

Cullen, P., Cahill, J. and Gaynor, K. (2021) 'A qualitative study exploring well-being and the potential impact of work-related stress among commercial airline pilots', *Aviation Psychology and Applied Human Factors*, 11(1), pp. 1–12. https://doi.org/10.1027/2192-0923/a000199

De Kock, J. H., Latham, H. A., Leslie, S. J., et al. (2021) 'A rapid review of the impact of COVID-19 on the mental health of healthcare workers: Implications for supporting psychological well-being', *BMC Public Health*, 21, p. 104. https://doi.org/10.1186/s12889-020-10070-3

Directorate General of Civil Aviation (DGCA) India (2023) *Medical Circular 01 of 2023: Mandatory Psychological Assessment and Peer Support Programme for Flight Crew*. New Delhi. Available at: https://www.dgca.gov.in/digigov-portal/?baseLocale=en_US?dynamicPage=dynamicPdf/KCvjSDbtfD3kCbIs9CgyYA%3D%3D&maincircularsRulesMedicalCirculars/7/4724/viewDynamicRuleContLvl2

Edmondson, A. C. (2018) *The Fearless Organization: Creating Psychological Safety in the Workplace for Learning, Innovation, and Growth*. Hoboken, NJ: Wiley.

Erickson, K. I., Voss, M. W., Prakash, R., et al. (2011) 'Exercise training increases size of hippocampus and improves memory', *Proceedings of the National Academy of Sciences*, 108(7), pp. 3017–3022. https://doi.org/10.1073/pnas.1015950108

European Union Aviation Safety Agency (EASA) (2016a) *ED Decision 2016/008/R: Acceptable Means of Compliance and Guidance Material to Part-ORO – Psychological Assessment and Support Programme Requirements*. Cologne: EASA. Available at: https://www.easa.europa.eu/en/document-library/agency-decisions/ed-decision-2016008r

European Union Aviation Safety Agency (2016b) *EASA Action Plan for Pilot Mental Fitness*. Cologne: EASA. Available at: https://www.easa.europa.eu/en/document-library/opinions/opinion-142016

European Union (2018) *Commission Regulation (EU) 2018/1042 of 23 July 2018 amending Regulation (EU) No 965/2012 as regards requirements for the psychological assessment of flight crew and alcohol testing*. Official Journal of the European Union, L 188, pp. 3–7. Available at: https://eur-lex.europa.eu/eli/reg/2018/1042/oj

European Union Aviation Safety Agency (EASA) (2023) *Peer Support Programmes (PSP) – AMC/GM updates/MeSAFE*. Available at: https://www.easa.europa.eu/en/domains/aircrew-and-medical/peer-support-programmes

Federal Aviation Administration (FAA) (2024) *Mental Health Aviation Rulemaking Committee (ARC) Final Report*. Washington, DC: FAA. Available at: https://www.faa.gov

Flight Safety Foundation (2024) *Path to Wellness: Charting a New Course for Mental Health in Aviation*. Alexandria, VA: FSF. Available at: https://flightsafety.org/wp-content/uploads/2024/10/FSF_MentalHealthWhitePaper_V5.pdf

Fredes-Collarte, D., Olivares-Faúndez, V., Sánchez-García, J. C. and Peralta Montecinos, J. (2024) 'Work satisfaction and its relationship with burnout and its consequences, using a structural model, in air cabin crew members', *Sustainability*, 16(22), p. 9619. https://doi.org/10.3390/su16229619

General Civil Aviation Authority (GCAA) (2024) *Psychoactive-Substance Management Program*. Dubai: United Arab Emirates. Available at: https://www.gcaa.gov.ae/en/epublication/NPADocuments/NOTICE%20OF%20PROPOSED%20AMENDMENT%20%202024-08%20CAR-PSMP%20-%20PSYCHOACTIVE%20SUBSTANCES%20MANAGEMENT%20PROGRAMME%20-%20ISSUE%2001.pdf

Goode, J. H. (2003) 'Are pilots at risk of accidents due to fatigue?', *Journal of Safety Research*, 34(3), pp. 309–313. https://doi.org/10.1016/S0022-4375(03)00033-1

Henderson, C., Evans-Lacko, S. and Thornicroft, G. (2013) 'Mental illness stigma, help seeking, and public health programs', *BMC Medicine*, 11, p. 80. https://doi.org/10.1186/1741-7015-11-80

Hoffman, W. R., Patel, P. K., Aden, J., Willis, A., Acker, J. P., Bjerke, E., Miranda, E., Luster, J. and Tvaryanas, A. (2023) 'Multinational comparison study of aircraft pilot healthcare avoidance behaviour', *Occupational medicine (Oxford, England)*, 73(7), pp. 434–438. https://doi.org/10.1093/occmed/kqad091

Hülsheger, U. R. and Schewe, A. F. (2011) 'On the costs and benefits of emotional labor: a meta-analysis of three decades of research', *Journal of Occupational Health Psychology*, 16(3), pp. 361–389. https://doi.org/10.1037/a0022876

ICAO (2022) *Manual on Human Performance for Regulators (Doc 10151)*. Montréal: International Civil Aviation Organization. Available at: *https://elibrary.icao.int/product/250419*

ICAO South America (SAM) Office (2023) *South American Regional Aviation Safety Plan (RASP) 2023–2025*. Lima: International Civil Aviation Organization, South America Office. Available at: https://www.icao.int/sites/default/files/safety/GASP/GASP%20Library/Regional%20aviation%20safety%20plans/SAMSP_2023-2025_en.pdf

International Air Transport Association (IATA) (2024) *Annual Safety Report – 2024 Recommendations for Accident Prevention*. Available at: https://www.iata.org/contentassets/95e933e1ad794068812f073cf883cb08/recommendations-for-accident-prevention-2024_final.pdf

Kalaga, L. (2023) 'Study of aviation psychology and its impact on human factors to substantially reduce human error through crew resource management', *International Journal of Scientific Development and Research (IJSDR)*, 8(2), pp. 373–377. Available at: https://www.ijsdr.org/papers/IJSDR2302066.pdf

Laborde, S., Mosley, E. and Thayer, J. F. (2017) 'Heart rate variability and cardiac vagal tone in psychophysiological research – recommendations for experiment planning, data analysis, and data reporting', *Frontiers in Psychology*, 8, p. 213. https://doi.org/10.3389/fpsyg.2017.00213

Lim, J. and Dinges, D. F. (2010) 'A meta-analysis of the impact of short-term sleep deprivation on cognitive variables', *Psychological Bulletin*, 136(3), pp. 375–389. https://doi.org/10.1037/a0018883

Littlepage, G. E., Hein, M. B., Moffett, R. G., Craig, P. A. and Georgiou, A. M. (2016) 'Team training for dynamic cross-functional teams in aviation: Behavioral, cognitive, and performance outcomes: Behavioral, cognitive, and performance outcomes', *Human Factors: The Journal of the Human Factors and Ergonomics Society*, 58(8), pp. 1275–1288. https://doi.org/10.1177/0018720816665200

Martinussen, M. and Hunter, D. R. (2017) *Aviation Psychology and Human Factors*. 2nd ed. Boca Raton, FL: CRC Press.

Mayday, S. A. (2025) *Peer-support MoUs/guidance*. Available at: https://mayday-sa.org.za/

Melin, M. and Lång, V. (2023) 'Are peer support programs effective in the detection and prevention of mental health issues in commercial aviation?', *The International Journal of Aerospace Psychology*, 34(3), pp. 162–175. https://doi.org/10.1080/24721840.2023.2278670

Miller, A. H. and Raison, C. L. (2016) 'The role of inflammation in depression: From evolutionary imperative to modern treatment target', *Nature Reviews Immunology*, 16(1), pp. 22–34. https://doi.org/10.1038/nri.2015.5

Nestler, E. J. and Carlezon, W. A. Jr (2006) 'The mesolimbic dopamine reward circuit in depression', *Biological Psychiatry*, 59(12), pp. 1151–1159. https://doi.org/10.1016/j.biopsych.2005.09.018

NTSB (2024) *Aviation Investigation Preliminary Report – Horizon Air/Alaska Airlines Flight 2059 (DCA24MA063)*. Washington, DC: National Transportation Safety Board. Available at: https://www.ntsb.gov/investigations/Documents/DCA24MA063%20Preliminary%20report.pdf

Olaganathan, R. and Amihan, R. A. H. (2021) 'Impact of COVID-19 on pilot proficiency – a risk analysis', *Global Journal of Engineering and Technology Advances*, 6(3), pp. 1–13. https://doi.org/10.30574/gjeta.2021.6.3.0023

South African Civil Aviation Authority (SACAA) (2023) *Mood Disorder Protocol for Aviation Medical Examiners*. Midrand: South African Civil Aviation Authority. Available at: https://www.caa.co.za/Documents/Medical/Mood_Disorder_Protocol.pdf

Thayer, J. F., Åhs, F., Fredrikson, M., Sollers, J. J. III and Wager, T. D. (2012) 'A meta-analysis of heart rate variability and neuroimaging studies: Implications for HRV as an index of top-down prefrontal control of subcortical regions', *Neuroscience & Biobehavioral Reviews*, 36(2), pp. 747–756. https://doi.org/10.1016/j.neubiorev.2011.11.009

Transport Canada (2024). *Psychological safety in aviation*. Aviation Safety Letter Issue 2/2024. Ottawa: Transport Canada. Available at: https://tc.canada.ca/en/aviation/publications/aviation-safety-letter/issue-2-2024/psychological-safety-aviation

Transport Canada (2025) *Mental health and crew resource management*. Aviation Safety Letter Issue 2/2025. Ottawa: Transport Canada. Available at: https://tc.canada.ca/en/aviation/publications/aviation-safety-letter/issue-2-2025/mental-health-crew-resource-management

Twenge, J. M., Cooper, A. B., Joiner, T. E., Duffy, M. E. and Binau, S. G. (2019) 'Age, period, and cohort trends in mood disorder indicators and suicide-related outcomes in a nationally representative dataset, 2005–2017', *Journal of Abnormal Psychology*, 128(2), pp. 119–128. https://doi.org/10.1037/abn0000410

UK Civil Aviation Authority (UK CAA) (2023) *Mental Health Guidance Material (MED.055)*. London. Available at: https://www.caa.co.uk/aeromedical-examiners/medical-standards/pilots/medical-conditions/mental-health/mental-health-guidance-material-gm/

WHO (2022) *World Mental Health Report: Transforming Mental Health for All*. Geneva: World Health Organization. Available at: https://www.who.int/publications/i/item/9789240049338

Wu, A. C., Donnelly-McLay, D., Weisskopf, M. G., Galligan, C. J., Porter, J. H., Wade, R. L. and Christiani, D. C. (2016) 'Depression symptoms among airline pilots', *Environmental Health*, 15, p. 121. https://doi.org/10.1186/s12940-016-0200-6

Yoo, S.-S., Gujar, N., Hu, P., Jolesz, F. A. and Walker, M. P. (2007) 'The human emotional brain without sleep – A prefrontal-amygdala disconnect', *Nature Neuroscience*, 10(3), pp. 377–379. https://doi.org/10.1038/nn1847

4 Stress and Resilience
Coping above the Clouds

CASE STUDY: US AIRWAYS FLIGHT 1549 (2009) – THE HUDSON RIVER LANDING

On January 15, 2009, US Airways Flight 1549 departed from New York's LaGuardia Airport bound for Charlotte, North Carolina. The Airbus A320 climbed into a cold winter sky carrying 150 passengers and five crew members, with two pilots on the flight deck: Captain Chesley "Sully" Sullenberger and First Officer Jeffrey Skiles. Though they had not flown together before, both were highly experienced. Sullenberger, a former US Air Force fighter pilot, had logged more than 19,000 flight hours and had extensive training in accident investigation and safety procedures. He had long been interested in human factors and the role of human performance in safety. Skiles had about 15,000 hours and had recently completed A320 simulator training that included engine-out scenarios (NTSB, 2010; Sullenberger and Zaslow, 2009).

At roughly 3,000 feet and about 100 seconds after takeoff, the aircraft encountered a flock of Canada geese. A loud thud was followed by an eerie quiet as both engines failed. The aircraft began to descend. In seconds, a routine departure turned into a time-critical emergency. Air traffic control asked for intentions. Sullenberger assessed options and replied with calm precision that they would be in the Hudson. With LaGuardia unreachable and Teterboro too far, Skiles worked the checklist while Sullenberger flew the jet, coordinating a glide approach over Manhattan. At 15:31, the A320 touched down on the Hudson River. The aircraft remained largely intact, doors opened quickly, rafts deployed, ferries converged, and every one of the 155 people on board survived without major injury (NTSB, 2010).

The press described it as a miracle. For the crew, it was something different: the product of training, discipline, and composure. Sullenberger later explained that he drew on everything he had ever learned and rehearsed, from procedural rigor to mental steadiness in the face of stress (Sullenberger and Zaslow, 2009). This was not just skilled flying. It was resilience under extreme pressure, enabled by long experience, effective crew resource management, and recent simulator practice that made the unthinkable seem manageable in the few seconds that mattered most.

DOI: 10.1201/9781003515418-4

INTRODUCTION

Stress and resilience in aviation are not abstract concepts. They are lived realities for different roles in the industry: pilots, cabin crew, controllers, or instructors. A flight that begins smoothly can change within minutes. Weather shifts. A technical system falters. A clear plan suddenly needs a backup. In those sharp, high-pressure moments, it is resilience that protects performance and keeps safety intact. Technology plays its part, but in the end, it is the human operator's ability to regulate stress and recover function quickly that determines how the story ends (Flin et al., 2008; Woods and Hollnagel, 2006).

Biologically, stress is not always negative. In fact, it is designed to help us survive. When a sudden challenge appears, the body responds with a cascade of hormones, most notably cortisol and adrenaline, which sharpen focus and accelerate action. For a short while, this can be useful. A pilot executing a go-around in poor weather may benefit from that burst of alertness. But stress that lasts too long, or arrives too intensely, begins to narrow thinking, fragment memory, and strain communication and teamwork (Leach, 2005; McEwen, 2007; Sapolsky, 2004).

Resilience is what balances this equation. It is the learned capacity to keep functioning effectively under strain, and to restore performance once the peak of stress has passed. It is not one single trait, but a mix of mindset, experience, and skills that can be deliberately developed. High-fidelity simulation, scenario-based training, and structured debriefs are some of the tools used to build it. They normalize stress, help aircrew practice adaptive responses, and anchor confidence in strategies that are already familiar when the pressure hits. First Officer Skiles' recent simulator sessions on engine-out procedures are a perfect example of how this preparation can transfer directly to real-world performance under extreme conditions (Kanki et al., 2010; NTSB, 2010).

Over the past decade, the way the industry talks about these issues has shifted. The focus is no longer only on error avoidance. It has widened to include human performance in its broadest sense: how people adapt, how they regulate their reactions, and how they prepare for the unexpected. Organizations such as International Civil Aviation Organization (ICAO), Federal Aviation Administration (FAA), or European Union Aviation Safety Agency (EASA) now position resilience as a core competency, essential for safe operations and not a bonus trait that some individuals happen to have (EASA, 2015; FAA, 2001; ICAO, 2013.

INDUSTRY SHIFTS AND ORGANIZATIONAL PRACTICES

Aviation's understanding of **stress and resilience** has evolved remarkably over the past four decades. Once treated as side topics to technical performance, they are now seen as defining features of professional competence. This shift was not sudden. It grew from accidents, advances in human factors science, and a deeper realization that even the most skilled crews operate within complex, high-stress systems. What has emerged is a more human-centered view of safety – one that recognizes performance not as perfect execution, but as the ability to regulate stress and recover function when pressure peaks.

From Error Prevention to Adaptive Performance

Early safety thinking assumed that if human error could be eliminated, safety would naturally improve. Training focused on procedures, compliance, and technical mastery. Yet investigations (from Tenerife to Sioux City) showed that technical skill alone could not explain why some crews held steady under pressure while others unraveled (Flin et al., 2008).

The conversation shifted from *avoiding error* to ***managing stress.*** Physiological stress responses – **elevated arousal, narrowed attention, fragmented memory** – became recognized as normal reactions, not personal flaws. The industry began to ask not only how to prevent errors, but how to prepare people to stay functional when stress is unavoidable. This was the foundation of adaptive performance: the ability to regulate arousal, make sense of ambiguous information, and stabilize behavior under pressure (Flin et al., 2008; Helmreich and Merritt, 2001; Kanki et al., 2010).

Resilience, in this light, became the second half of the stress equation – the active recovery process that restores bandwidth after the physiological surge. Training and policy began to reflect this cycle: recognize stress, regulate it, and rebuild performance (Hollnagel et al., 2011).

From Accidents to Resilience: A Timeline of Shifting Focus

The evolution of how aviation manages stress and resilience has been shaped less by academic theory and more by hard experience. Every major event, tragic or triumphant, has taught new lessons about human limits and adaptability. Figure 4.1 summarizes these turning points. It begins with Tenerife, where stress and communication failure combined fatally, and continues through events like United 232 at Sioux City (where crew improvisation under catastrophic failure highlighted resilience), US Airways 1549 on the Hudson, and Qantas 32 in Singapore, which demonstrated how training and composure can transform crisis into survival. Later milestones such as Germanwings 9525 and the global disruption of COVID-19 shifted focus toward organizational stress, peer support, and well-being (MeSAFE Consortium, 2024; Helmreich and Merritt, 2001; Kanki et al., 2010).

The figure is not a count of accidents but a visual story of progress – from **Safety-I**, focused on preventing error, to **Safety-II** (discussed in details in the chapter about safety culture) focused on reinforcing what allows people to succeed under pressure (Hollnagel, 2015). It captures the growing recognition that stress cannot be eliminated but can be managed and that resilience emerges when organizations learn to turn physiological strain into psychological readiness.

Training, Policy, and People – Embedding Resilience in Practice

Supporting this evolution is a broader human-performance ecosystem that aims to detect, manage, and learn from stress before it compromises safety. Airlines now embed structured **peer support**, **well-being monitoring**, and **psychological safety**

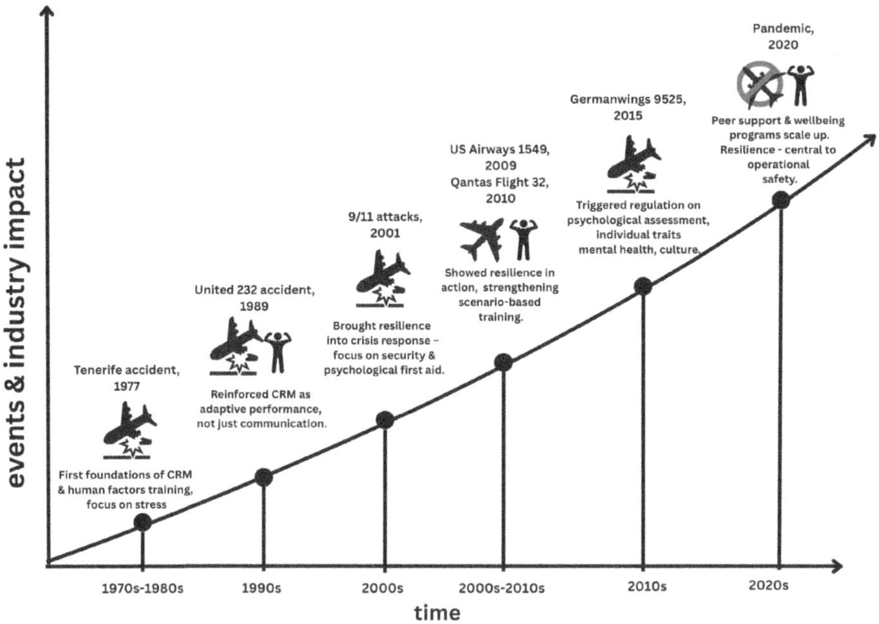

FIGURE 4.1 Key events timeline – how major accidents shaped resilience thinking.

practices into daily operations – not as emergency tools, but as preventive measures. The goal is early recovery, so keeping cognitive and emotional resources available before overload turns into error (Dekker, 2012; ICAO, 2013).

Training programs increasingly combine **stress inoculation** and **scenario realism**, exposing crews to incomplete data, degraded automation, and time-critical decision-making. Instructors are trained to observe physiological and behavioral signs of strain: voice tension, loss of verbal coordination, or narrowing attention, and use structured debriefs to discuss recovery strategies. In this way, stress management becomes as coachable as procedural skill.

Aviation psychologists, well-being specialists, and qualified pilots trained in human factors now collaborate on evidence-based programs that translate stress science into practice. They contribute to **Competency-Based Training and Assessment (CBTA) and Evidence-Based Training (EBT) frameworks** by defining behavioral markers for self-regulation, monitoring, and composure. The aim is to treat stress regulation as part of normal competency, not a separate domain reserved for clinical attention (EASA, 2015; ICAO, 2013 IATA, 2024).

Since the Germanwings 9525 tragedy, regulators such as **EASA** have formalized this connection by mandating psychological assessments, ongoing monitoring, and organizational systems that view mental health and stress tolerance as core components of operational safety (EASA, 2016). Together, these initiatives reflect a more mature understanding: resilience is not a trait that appears after stress, but a capacity built through continuous management of it.

Resilience across The Roles

The ability to manage stress under pressure is not confined to the flight deck. Cabin crew, dispatchers, and maintenance engineers all work in time-sensitive, high-stakes environments, each facing a unique blend of acute and cumulative stressors – from disruptive passengers to weather delays and mechanical troubleshooting under operational pressure. Our understanding of stress and resilience across these roles has also evolved (or maybe even more); we now recognize that managing stress is an inherently **cross-functional and interconnected process**, rather than an individual skill exercised in isolation (Kanki et al., 2010)

Training programs now integrate **cross-role scenarios** that bring these stressors into shared learning contexts: cockpit-cabin coordination during emergencies, maintenance-ops communication under tight turnarounds, or joint decision-making in irregular operations. These exercises strengthen both **individual stress awareness** and **team resilience** (Littlepage et al., 2016).

Organizations also analyze debriefs, fatigue reports, and Crew Resource Management (CRM) assessments for early signs of overload, adapting schedules, briefings, and procedures to reduce unnecessary stress. By developing a shared behavioral language for stress and recovery, they make it possible to coach teams systemically rather than reactively (IATA, 2024; Kanki et al., 2010).

LEVELS OF STRESS IN AVIATION

Stress is not one uniform thing. It comes in different shapes, lasts for different durations, and affects performance in different ways. Breaking it down makes it easier to see why sometimes stress sharpens a pilot's focus, and at other times it unravels decision-making altogether. Thinking in terms of impact, source, and duration gives us a practical framework for how to recognize stress in the cockpit or cabin, and how to deal with it.

Impact on Performance

The first lens is what stress does to performance. Short bursts can sharpen the scan, tighten focus, and speed up execution – the kind of stress response that helps a captain pull off a tight go-around. That is what psychologists call facilitative stress, or eustress. But if arousal goes too high, or stays high too long, it overwhelms cognitive capacity. Memory fragments, communication falters, and decision-making slides (Folkman, 2013).

- **Facilitative stress (eustress)** – sharpens focus and execution for a short time.
- **Debilitative stress (distress)** – overwhelms cognitive resources and degrades decision-making (Folkman, 2013).

Type of Stressors

Different sources of stress load different systems. Environmental and operational stressors often tax sensory processing and workload management, while interpersonal stressors drain attentional control and communication bandwidth. Knowing the source helps target mitigation, for example through checklist discipline, crew coordination, or environmental controls (Cooper et al., 1988).

- **Environmental** – noise, temperature extremes, turbulence, circadian disruption.
- **Operational –** delays, system malfunctions, schedule variability.
- **Interpersonal** – cockpit dynamics, passenger aggression, unclear Air Traffic Control (ATC) instructions (Cooper et al., 1988).

Duration of Stress

Finally, stress can also be understood by how long it lasts.

- **Acute stress** is the sharp spike (sudden, high-intensity events) – a sudden decompression in the cabin, a disruptive passenger turning aggressive, or an unexpected systems alert. It demands quick regulation to keep us available for action.
- **Chronic stress** is quieter but more corrosive – endless roster changes, irregular sleep from night shifts, or the constant pressure of short turnaround times. Over weeks or months, it slowly erodes energy, mood, and performance.
- **Cumulative stress** is different again. It is the build-up of many small hits: a string of minor delays, missed breaks, or repeated last-minute gate changes (Contrada and Baum, 2011).

Each one on its own is tolerable, but without recovery time they add up, often surfacing as fatigue, irritability, or lapses in attention (McEwen, 2007). Transactional models remind us that stress is not only about the event itself, but about how it is interpreted and whether coping resources are available (Folkman, 2013). That is why the same thunderstorm diversion can feel manageable to one crew and overwhelming to another. Flight 1549 illustrates how acute technical failure and time pressure can be absorbed and managed if resilience is strong enough. By contrast, as seen in accidents like Air France 447, stress without sufficient resilience or experience leaves little margin for recovery (BEA, 2012; NTSB, 2010).

KEY COMPONENTS OF RESILIENCE

Resilience in aviation isn't a mysterious single trait that some people are lucky to be born with. It's a collection of skills and habits that can be observed, trained, and strengthened over time. What keeps a crew safe under pressure is not a single quality but a mix of mental and social capacities that work together. Resilience is like a pathway built from multiple capacities – each waypoint strengthens the ability to cope, recover, and perform under stress.

1. **Emotional regulation** is the first. The ability to manage anxiety, frustration, or even fear keeps working memory clear enough for sound judgment. Crews who can steady their breathing, keep their voice calm, and focus on the task preserve the mental space needed to make decisions (Gross, 1998).

2. **Self-efficacy** is another key element. Believing you can influence events – even when circumstances are tough – supports initiative under stress. A pilot with high self-efficacy is more likely to take decisive action instead of freezing, and a cabin crew member with the same confidence will step in firmly to de-escalate conflict rather than avoid it (Bandura, 1997).
3. **Cognitive flexibility** matters too. Stress often comes with uncertainty and shifting cues. Being able to adapt your mental model, drop a plan that no longer works, and shift strategies as the situation evolves is a hallmark of resilient performance (Scott, 1962).
4. **Social support** is the buffer that keeps individuals from carrying the entire load alone. Reliable communication, trust within a crew, and confidence that colleagues will back you up all help regulate stress. In practice, this might mean a first officer voicing an observation just in time, or a cabin crew colleague quietly stepping in to calm a tense passenger. These small actions can change the trajectory of a stressful situation (Flin et al., 2008).
5. Finally, **experience-based confidence** ties the components together. Repetition and exposure ground actions in known strategies. Each successful response to turbulence, conflict, or technical malfunctions adds to the mental library that can be drawn on when a fresh crisis appears. That stored experience makes composure under stress easier to access when time is short (Salas et al., 2012).

Taken together, these components show that resilience is not fixed – it's built. Through deliberate practice, honest feedback, and organizational support, aviation professionals can strengthen these capacities so that when the real test comes, they have the tools to hold steady.

THE SCIENCE BEHIND STRESS AND RESILIENCE: PERSONALITY, COMPETENCIES, AND EXPERIENCE

Industry practice shows us *what* stress and resilience looks like in the cockpit, the cabin, or the ops room. The science helps us understand *why* some people cope better than others, and how those abilities can be trained. Stress is, at its core, the body's adaptive alarm. It mobilizes attention and action, but when it continues for too long it eats away at memory, decision-making, and coordination. In aviation, short bursts of stress can sharpen focus just enough to get a go-around done safely. But chronic load, or stress that never gets a reset, gradually erodes decision quality and team communication (Folkman, 2013; Leach, 2005; McEwen, 2007; Sapolsky, 2004).

Resilience is the counterbalance. It is not about being immune to stress, but about being able to take the hit, recover quickly, and stay functional when the demands are high. And it is not fixed – it can be built through exposure, reflection, and practice. That is why training that includes graded stress, structured debriefing, and feedback loops is so valuable: it normalizes pressure, rehearses coping, and builds confidence in strategies that can hold up when reality bites (Fletcher and Sarkar, 2013; Kanki et al., 2010; Robertson et al., 2015).

Individual Differences (Personality)

No two pilots or crew members react to stress in exactly the same way. Some stay calm and methodical, others tighten up or go quiet. Personality matters here. Traits like emotional stability, conscientiousness, or openness tend to give people a steadier hand when the workload spikes. Optimism, self-efficacy, and adaptability also help people engage with problems actively instead of retreating or freezing (Chen et al., 2025; Costa and McCrae, 1992).

Research has shown that some personality traits correlate more strongly with pilot performance than others, although findings vary depending on study design, sampling, and the tools used (Cox et al., 2000). The aim here, however, is not to debate methodological precision, but to apply these insights constructively – to recognize behavioral patterns that, in operational settings, may increase susceptibility to risk, complacency, or reduced adaptability, and to use this understanding to strengthen selection and training practices (EASA, 2016). Professional personality assessments designed for aviation field can give an early picture of how someone is likely to hold up under pressure (Chen et al., 2025). Pair that with simulator runs that test judgment in real time, and you get a sharper sense of not only what a trainee knows, but how they behave when the stress hits. That knowledge lets instructors and training organizations shape development before weak points become safety risks (Cox et al., 2000; Flin et al., 2008).

In that way, simulator-based cognitive measures add another layer, showing how personality translates into behavior when workload rises, being a great tool to adjust the pressure to create positive learning environment. Together, these implementations allow training providers to tailor development plans to individual needs, addressing vulnerabilities before they surface on the line (Flin et al., 2008).

There are other individual factors that play an important role in reacting and responding to stress, like adaptability, cognition, or learning abilities. Resilience doesn't begin in procedures – it begins in how the individual system processes the load. Once again, we need to approach this as a complicated ecosystem, which starts to work only when treated holistically – when inner and outer systems communicate (see Chapter 1, the SPINAL-CORD model).

Competencies

Resilience is not just about who you are, but also about what you can do. And in aviation (if the level of individual traits allows), that comes down to visible, trainable skills. Can you regulate your emotions enough to keep thinking straight? Can you switch mental gears when the plan collapses? Can you share the load with your team without letting communication fray? Those are the competencies that matter.

This is where training methods make the difference. Stress inoculation sessions push crews into pressure while teaching short, usable coping moves. Cognitive rehearsal and mental practice make memory items and decision steps faster to access. Structured debriefs then tie it all together by showing which behaviors held up and which cracked under strain (Flin et al., 2008; Meichenbaum, 2007; Salas et al., 2012). The evidence is clear: when these competencies are deliberately drilled, errors fall in simulations and the gains carry over into live flying (Fletcher and Sarkar, 2013; Kanki et al., 2010; Robertson et al., 2015).

Experience

Experience is the piece that ties everything together. It teaches you not just the checklist, but the feel – when something is sliding off track, when a cue matters, when you need to slow down before the situation runs away from you. With enough varied exposure, crews learn to spot the danger signs earlier, choose workable actions faster, and recover mental space after a heavy workload. Instructors play a big role too: noticing when a student is overloaded, stepping in without killing the learning, and building tolerance in layers (Kanki et al., 2010; NTSB, 2010; Salas et al., 2012). And when that experience is thin, the cracks show.

A prime example of this is **Air France Flight 447** in June 2009. When pitot tubes iced over and airspeed data became unreliable, the crew of the A330 encountered startle, confusion, and conflicting information at night over the Atlantic. Under stress, and with little prior experience of such high-altitude failures, their resilience eroded. Manual flying deteriorated, and the aircraft entered a stall from which it never recovered, crashing into the ocean and killing all 228 people on board. The Bureau d'Enquêtes et d'Analyses (BEA) investigation later showed how a lack of exposure to ambiguous, high-altitude events left the crew without the mental models or practiced responses needed to adapt. It remains a stark reminder that resilience is not theoretical, it must be built through structured practice before stress strips it away (BEA, 2012).

INDUSTRY SHIFTS AND ORGANIZATIONAL PRACTICES

Aviation's relationship with stress has evolved dramatically over the past four decades. Once viewed as an individual issue, a matter of personal endurance or temperament, it is now understood as a predictable element of complex operations. The question is no longer whether stress will appear, but how well the system prepares people to handle it. Training, regulation, and support structures have gradually expanded from managing emergencies to sustaining performance in everyday pressure. This change reflects a deeper cultural shift: from reacting to stress to deliberately regulating it across teams, roles, and organizations.

Managing Stress: From Reaction to Regulation

Early safety systems focused on preventing errors through technical precision and procedural compliance. Yet, as accidents like Tenerife in 1977 and United 232 in 1989 showed, technical skill alone could not safeguard performance when cognitive overload or emotional stress took hold. Investigations revealed that even highly experienced crews could lose coordination and situational awareness once physiological arousal crossed a threshold (ICAO, 1978; NTSB, 1990).

Modern human factors research reframed this problem. Stress was no longer seen as a sign of weakness, but as a **biological response to challenge** – a temporary surge of cortisol and adrenaline that narrows attention, speeds reaction, and, if unchecked, distorts reasoning. The industry began to study how stress influences communication, decision-making, and recovery. This shift moved safety thinking from error avoidance to **adaptive regulation**, so teaching crews how to recognize and manage the effects of stress before they degrade performance (Flin et al., 2008; Helmreich and Merritt, 2001).

By the late 1990s, CRM evolved into a broader system of stress awareness and coping strategies. Rather than trying to remove pressure entirely, training began to simulate it in controlled doses, allowing crews to build tolerance and confidence under realistic workloads. This approach laid the foundation for the industry's current resilience frameworks.

Embedding Stress Awareness in Training and Policy

Today, managing stress is no longer limited to classroom discussions or simulator briefings – it is embedded across aviation operations. Training programs integrate **stress inoculation**, **scenario realism**, and **behavioral observation**, ensuring that both technical and non-technical competencies are assessed under time pressure and uncertainty. Instructors are trained to look for subtle cues of overload: clipped speech, rigid scanning, or rushed checklists. Structured debriefs then address not only procedural accuracy but also how crews regulated their own stress responses.

Supporting these efforts is a broader ecosystem of **peer support**, **well-being monitoring**, and **psychological safety practices**. These systems help detect both acute and chronic stress before they compromise performance. The goal is proactive regulation – maintaining cognitive bandwidth and emotional stability rather than waiting for burnout or breakdown (Dekker, 2012; IATA, 2024; ICAO, 2013).

Aviation psychologists, human factors specialists, and qualified pilot-trainers now collaborate on program design, behavioral markers, and coaching techniques that translate stress science into daily practice. Within **CBTA** and **EBT** frameworks, stress management is treated as a measurable element of professional competence. Since the Germanwings 9525 tragedy, regulators such as EASA have further emphasized this connection by mandating psychological assessment, ongoing monitoring, and organizational systems that view mental health and stress tolerance as core components of operational safety (EASA, 2016; IATA, 2024).

The outcome is a more mature model: one in which stress regulation is not an afterthought or recovery skill but an active, continuous process – built through design, culture, and deliberate training.

Lessons from the Field: Stress, Recovery, and Growth

The industry's understanding of stress has been shaped as much by experience as by research. Each major event (whether tragedy or success) has offered new insight into how humans react under intense physiological and emotional strain, and how communication and leadership can steady performance in chaos.

Let's look at the so-called washing machine incident. On June 25, 2017, **AirAsia X Flight D7237** departed Perth for Kuala Lumpur with 357 passengers and 10 crew members. About two hours into the flight, the Airbus A330-343 (9M-XXE) experienced a catastrophic failure of its left engine, producing violent vibration that shook the entire airframe. The cockpit was flooded with noise and warning messages, while passengers later described the experience as "like being inside a washing machine."

The captain and first officer faced both acute physical stress and extreme workload pressure. Despite the sensory overload, the captain maintained deliberate composure – stabilizing the flight path, delegating tasks clearly, and communicating openly with the cabin. His calm yet authoritative tone acknowledged the seriousness of the event without fueling panic. After nearly 90 minutes of continuous vibration, the crew landed safely back in Perth

Investigators later noted that the crew's calm communication and deliberate pacing prevented panic among passengers and preserved cockpit coordination (ATSB, 2019).

For aviation psychology, the incident demonstrated that stress management is not only internal control but also **emotional influence** – how a leader's demeanor can stabilize others in a shared crisis. It remains a vivid example of adaptive performance under physiological and cognitive load.

POSITIVE DEVELOPMENTS FOR NOW AND THE FUTURE

US Airways Flight 1549 reframed public and industry attention on what goes **right**. It showed that composure, teamwork, and training can transform crisis into control. In the years since, regulators and airlines have increasingly embraced positive performance psychology – a shift from preventing errors to reinforcing the behaviors and environments that enable success under stress.

IATA's *Human Performance in Aviation* guidance emphasizes evidence-based, scenario-driven training approaches that expose crews to **ambiguity, prioritization demands, and adaptive decision-making** alongside technical tasks (IATA, 2024) In parallel, several major international airlines have begun integrating **resilience and stress-management elements** into CRM refreshers and recurrent simulator training. These initiatives increasingly focus on emotion regulation, physiological awareness, and recovery from stress – capabilities once viewed as intangible, but now approached as observable and trainable behaviors within human performance frameworks (He et al., 2025).

Expanding Frontiers of Stress Science

Recent advances in neurocognitive and physiological monitoring are transforming how stress is studied and coached. While simulator fidelity continues to improve, the breakthrough lies in real-time data. Technologies such as **eye-tracking, heart rate**

variability (HRV), galvanic skin response, and pupil dilation analysis now provide objective indicators of attention, arousal, and workload – metrics directly tied to stress response. For example, machine learning models combining HRV features have achieved high level accuracy (over 80%) in classifying pilot workload in simulated low-visibility conditions (Wang et al., 2025), and live flight tests have successfully merged eye tracking, heart rate, and brain imaging data to monitor cognitive engagement (Monk et al., 2024).

Used carefully, these tools don't reduce resilience to a number, but contextualize it. Instructors can identify when a pilot's attention narrows excessively, when physiological arousal climbs too high, or when micro-pauses in communication signal overload. When combined with post-scenario debriefs, these markers allow stress to be **seen, discussed, and managed** rather than guessed. The integration of biofeedback into advanced training also helps pilots and instructors recognize their own physiological signatures of stress and apply short regulation techniques – controlled breathing, cognitive reframing, and micro-recovery pauses – proven to restore situational awareness (Robertson et al., 2015; Saus et al., 2012).

Building Supportive Systems

The evolution of training has been matched by cultural and organizational change. Airlines increasingly treat **stress management as a shared responsibility** rather than a personal challenge. Peer-support programs have matured from reactive "after-incident" services into preventive systems embedded within Safety Management Systems (SMS). Many operators now link fatigue reporting, psychological support, and operational data to identify early signs of chronic stress at team or fleet level (AsMA 2024; MeSAFE Consortium, 2024).

Leaders are also being trained to notice subtle indicators of overload in their teams: irritability, loss of humor, withdrawal, and to protect recovery time with the same rigor used for technical compliance. Instructors and check pilots are encouraged to model openness about stress, making regulation and recovery legitimate topics of debrief rather than taboo subjects. This normalization of **stress talk** marks a cultural shift: resilience is no longer stoicism, but intelligent energy management.

As this chapter closes, carry forward one idea: **stress management and resilience is not luck – it's deliberate design.** It emerges from the intersection of training, culture, and science.

TAKEAWAYS

For Pilots and Cabin Crew

- Recognize your stress signals early: tight scanning, silence, or irritability are cues to pause and reset, not push harder.
- Use short regulation techniques such as controlled breathing, mental checklists, or intent statements to restore focus during high workload moments.

- Treat recovery as part of your job. Small resets between sectors or flights sustain attention and decision quality more effectively than pushing through fatigue.

Reflective Question: *When I notice stress building, do I act on it – or let it act on me?*

For Instructors, Examiners, and Leaders

- Observe how people *stabilize* under stress, not just how they perform tasks. Coaching should focus on how awareness and communication are preserved.
- Incorporate ambiguity and time pressure into scenario design to help trainees practice adaptability and composure, not just checklist execution.
- Model openness about stress. A leader who admits strain without losing clarity shows that composure is learned, not innate.

Reflective Question: *How do I use my position to make stress management visible and coachable rather than hidden and stigmatized?*

For Operational Teams

- Integrate stress monitoring into existing safety data: combine fatigue reports, simulator metrics, and feedback from peer-support programs to identify emerging stress patterns.
- Treat stress management as a shared system function. Build policies that protect recovery time and reinforce psychological safety as part of SMS performance.
- Leverage technology: biofeedback, eye-tracking, HRV monitoring, not for surveillance but to understand and coach human performance under load.

Reflective Question: *How clearly does our organization show that managing stress is part of operational excellence, not a personal problem?*

REFERENCES

Aerospace Medical Association (AsMA) (2024) *Resolution 2024-01: Integrating a Program of Mental Health and Wellness into Safety Management Systems across Aviation*. Alexandria, VA: Aerospace Medical Association. Available at: https://asma.org/wp-content/uploads/2025/08/AsMA-Resolution-2024-01-Integrating-a-Program-of-Mental-Health-and-Wellness-into-Safety-Management-Systems-Across-Aviation.pdf

Australian Transport Safety Bureau (ATSB) (2019) *Engine Failure and Return to Perth Involving Airbus A330-343, 9M-XXE, AirAsia X D7237, 25 June 2017*. Canberra:

ATSB. Available at: https://www.atsb.gov.au/publications/investigation_reports/2017/aair/ao-2017-070

Bandura, A. (1997) *Self-Efficacy: The Exercise of Control.* New York: W H Freeman/Times Books/Henry Holt & Co.

Bureau d'Enquêtes et d'Analyses (BEA) (2012) *Accident on 1st June 2009 to the Airbus A330-203 F-GZCP, Air France Flight AF447 – Final Report.* Paris: BEA. https://bea.aero/fileadmin/documents/docspa/2009/f-cp090601.en/pdf/f-cp090601.en.pdf

Chen, J., Landman, A., Stroosma, O. and van Paassen, M. M. (2025) 'The effect of personality traits and flight experience on pilots' cognitive and affective responses to simulated in-flight hazards', *Aviation Psychology and Applied Human Factors*, 14(2), pp. 104–113. https://doi.org/10.1027/2192-0923/a000283

Contrada, R. J. and Baum, A.(Eds.) (2011) *The Handbook of Stress Science: Biology, Psychology, and Health.* New York: Springer Publishing Company.

Cooper, C. L., Sloan, S. J. and Williams, S. (1988) *Occupational Stress Indicator.* Windsor: NFER-Nelson.

Costa, P. T. Jr. and McCrae, R. R. (1992) *NEO PI-R: Professional Manual.* Odessa, FL: Psychological Assessment Resources.

Cox, T., Griffiths, A. J. and Rial-Gonzalez, E. (2000) *Research on Work-Related Stress. Report to the European Agency for Safety and Health at Work.* Luxembourg: Office for Official Publications of the European Communities. http://osha.europa.eu/en/publications/reports/203/view

Dekker, S. (2012) *Just Culture: Balancing Safety and Accountability.* 2nd ed. Boca Raton, FL: CRC Press. https://doi.org/10.4324/9781315251271

European Union Aviation Safety Agency (EASA) (2015) *Annex II to ED Decision 2015-022-R: Acceptable Means of Compliance and Guidance Material to Part-FCL (Amendment 1 – Resilience Development, Human Performance and Threat and Error Management).* Cologne: EASA. Available at: https://www.easa.europa.eu/sites/default/files/dfu/Annex%20II%20to%20ED%20Decision%202015-022-R%20(correction%20on%20p.%2026).pdf

European Union Aviation Safety Agency (EASA) (2016) *ED Decision 2016/008/R: Acceptable Means of Compliance and Guidance Material to Part-ORO – Psychological Assessment and Support Programme Requirements.* Cologne: EASA. Available at: https://www.easa.europa.eu/en/document-library/agency-decisions/ed-decision-2016008r

Federal Aviation Administration (FAA) (2001) *Advisory Circular 120-51E: Crew Resource Management Training.* Washington, DC: US Department of Transportation, Federal Aviation Administration. Available at: https://www.faa.gov/documentLibrary/media/Advisory_Circular/AC_120-51E.pdf

Fletcher, D. and Sarkar, M. (2013) 'Psychological resilience: A review and critique', *European Psychologist*, 18(1), pp. 12–23. https://doi.org/10.1027/1016-9040/a000124

Flin, R., O'Connor, P. and Crichton, M. (2008) *Safety at the Sharp End: A Guide to Non-Technical Skills.* Boca Raton, FL: CRC Press. https://doi.org/10.1201/9781315607467

Folkman, S. (2013) 'Stress: Appraisal and coping', in M.D. Gellman, J.R. Turner (Eds.), *Encyclopedia of Behavioral Medicine.* New York: Springer. https://doi.org/10.1007/978-1-4419-1005-9_215

Gross, J. J. (1998) 'The emerging field of emotion regulation: An integrative review', *Review of General Psychology*, 2(3), pp. 271–299. https://doi.org/10.1037/1089-2680.2.3.271

He, P., Lu, F., Sun, R., et al. (2025) 'Examining influencing factors and their hierarchical relationships in flight crew resilient behavior through a hybrid ISM-DEMATEL approach', *Scientific Reports*, 15, p. 1606. https://doi.org/10.1038/s41598-025-85990-4

Helmreich, R. L. and Merritt, A. C. (2001) *Culture at Work in Aviation and Medicine: National, Organizational and Professional Influences.* 1st ed. Abingdon: Routledge. https://doi.org/10.4324/9781315258690

Hollnagel, E. (2015) *Safety-I and Safety-II: The Past and Future of Safety Management.* London: NHS England. Available at: https://www.england.nhs.uk/signuptosafety/wp-content/uploads/sites/16/2015/10/safety-1-safety-2-whte-papr.pdf

Hollnagel, E., Paries, J., Woods, D. and Wreathall, J. (2011) *Resilience Engineering in Practice: A Guidebook.* Farnham: Ashgate.

International Air Transport Association (IATA) (2024) *Recommendations for Accident Prevention: Human Performance in Aviation.* Montreal: IATA. Available at: https://www.iata.org/contentassets/95e933e1ad794068812f073cf883cb08/recommendations-for-accident-prevention-2024_final.pdf

International Civil Aviation Organization (ICAO). (2013) *Manual on Evidence-Based Training (EBT).* Montreal: ICAO. (Published as Doc 9995)

Kanki, B. G., Helmreich, R. L. and Anca, J. (2010) *Crew Resource Management.* 2nd ed. Cambridge, MA: Academic Press. https://doi.org/10.1016/C2009-0-02518-5

Leach, J. (2005) 'Cognitive paralysis in an emergency: The role of the supervisory attentional system', *Aviation, Space, and Environmental Medicine*, 76(2), pp. 134–136. Available at: https://pubmed.ncbi.nlm.nih.gov/15732750/

Littlepage, G. E., Hein, M. B., Moffett, R. G. and Craig, P. A. (2016) 'Team training for dynamic cross-functional teams in aviation: Behavioural, cognitive, and performance outcomes', *Human Factors: The Journal of the Human Factors and Ergonomics Society*, 58(8), pp. 1275–1289. https://doi.org/10.1177/0018720816665200

McEwen, B. S. (2007) 'Physiology and neurobiology of stress and adaptation: Central role of the brain', *Physiological Reviews*, 87(3), pp. 873–904. https://doi.org/10.1152/physrev.00041.2006

Meichenbaum, D. (2007) 'Stress inoculation training: A preventative and treatment approach', in P. M. Lehrer, R. L. Woolfolk, W. E. Sime (Eds.), *Principles and Practice of Stress Management* (3rd ed., pp. 497–516). New York: The Guilford Press.

MeSAFE Consortium (2024) *MeSAFE Project: Deliverables D5.1 and D6.1 Reports.* Cologne: European Union Aviation Safety Agency. Available at. Available at: https://www.easa.europa.eu/en/research-projects/mesafe-mental-health#group-downloads

Monk, K. J., Duong, T., Pankok, C., Diaz-Artiles, A. and Kalar, K. (2024) Physiological and subjective responses of pilots during flight test evaluation. *NASA Technical Publication TP-2024-0007669.* Ames Research Center, Human Systems Integration Division. Available at: https://hsi.arc.nasa.gov/publications/NASATP_20240007669_IASFlightTest_HFreport.pdf

National Transportation Safety Board (NTSB) (1990) *Aircraft Accident Report: United Airlines Flight 232, McDonnell Douglas DC-10-10, Sioux Gateway Airport, Sioux City, Iowa, July 19, 1989.* Washington, DC: NTSB. Available at: https://www.ntsb.gov/investigations/AccidentReports/Reports/AAR-90-06.pdf

National Transportation Safety Board (NTSB) (2010) *Loss of Thrust in Both Engines after Encountering a Flock of Birds and Subsequent Ditching on the Hudson River, US Airways Flight 1549 (NTSB/AAR-10/03).* Washington, DC: NTSB. https://www.ntsb.gov/investigations/AccidentReports/Reports/AAR1003.pdf

Robertson, I. T., Cooper, C. L., Sarkar, M. and Curran, T. (2015) 'Resilience training in the workplace: From 2003 to 2014', *Journal of Occupational and Organizational Psychology*, 88(3), pp. 533–562. https://doi.org/10.1111/joop.12120

Salas, E., Tannenbaum, S. I., Kraiger, K. and Smith-Jentsch, K. A. (2012) 'The science of training and development in organizations', *Psychological Science in the Public Interest*, 13(2), pp. 74–101. https://doi.org/10.1177/1529100612436661

Sapolsky, R. M. (2004) *Why Zebras Don't Get Ulcers.* 3rd ed. New York: Holt. https://us.macmillan.com/books/9780805073690/whyzebrasdontgetulcers

Saus, E. R., Johnsen, B. H., Eid, J., Riisem, P. K., Andersen, R. and Thayer, J. F. (2012) 'The effect of brief situational awareness training in a simulated police shooting scenario', *Military Psychology*, 24(6), 544–556. https://doi.org/10.1207/s15327876mp1803s_2

Scott, W. A. (1962) 'Cognitive complexity and cognitive flexibility', *Sociometry*, 25(4), pp. 405–414. https://doi.org/10.2307/2785779

Sullenberger, C. B. and Zaslow, J. (2009) *Highest Duty: My Search for What Really Matters*. New York: HarperCollins.

Wang, Y., Li, J., Xu, H., Gao, Y., Zhao, L. and Li, H. (2025) 'Machine learning evaluation model of pilot workload in a simulated environment combining HRV features and subjective ratings', *Scientific Reports*, 15(1), p. 1179. Nature Portfolio. https://doi.org/10.1038/s41598-025-05759-7

Woods, D. D. and Hollnagel, E. (2006) *Resilience Engineering: Concepts and Precepts*. Boca Raton, FL: CRC Press. https://doi.org/10.1201/9781315605685

5 Battle for Alertness *Attention, Fatigue, and Sleep*

CASE STUDY: FLAIR AIRLINES FLIGHT 501 (2023) – HARD LANDING

It was still early when Flair Airlines Flight 501 lined up for arrival into Kitchener-Waterloo on February 9, 2023. The Boeing 737-800 had cruised uneventfully across Canada from Vancouver. Conditions were good: cold, clear, calm. One of those winter mornings when the crisp air seems to sharpen every edge. But not inside the cockpit.

On final to Runway 29, something shifted. At approximately 149 feet AGL, the aircraft suddenly pitched nose-down and began a rapid descent. No technical fault announced itself, no windshear warning, no bird strike – just an abrupt deviation during the final segment of the approach.

Moments later, the aircraft struck the runway nose-first. The nose gear collapsed, the forward fuselage absorbed the impact, alarms sounded, and passengers were injured as the aircraft came to a stop off the paved surface. Emergency slides were deployed. Everyone survived, but several people sustained injuries and the airframe suffered substantial damage (TSB, 2024).

The investigation established that the aircraft systems and automation were functioning normally throughout the approach; the event unfolded during manual flight. Flight data showed increasing deviations from a stabilized approach profile that were not effectively detected or corrected. Late recognition of changes in pitch attitude and descent rate was evident, with limited corrective control input before touchdown. Monitoring and callouts were inconsistently applied, and standard stabilization cues were not clearly established during the final approach segment. Investigators identified indicators of reduced situational awareness and degraded task management during this phase, consistent with fatigue-related performance impairment identified earlier in the flight (TSB, 2024).

Investigators ultimately turned away from mechanical failure and toward a quieter contributor: fatigue. The captain had accumulated sleep debt in the days prior, and reduced alertness may have affected monitoring and task management during the approach. Briefings were rushed, monitoring degraded, and callouts delayed. Neither pilot clearly recalled parts of the descent (TSB, 2024). As fatigued crews sometimes put it, they had "lost the thread."

Crew Resource Management (CRM) was present but muted. Automation was available, but it was not the limiting factor in this event. The human safety

DOI: 10.1201/9781003515418-5

net – monitoring, cross-checking, and timely intervention – was the final safeguard, and it was frayed. Flair 501 serves as a reminder that even in an environment of reliable systems and advanced avionics, fatigue can quietly erode the human performance margins upon which safe outcomes still depend.

INTRODUCTION: WHY ALERTNESS IS YOUR REAL CO-PILOT

Walk into any modern cockpit and you're greeted by systems designed to make flying safer and smoother – layers of automation, digital checklists, thoughtful human–machine interface design. The airplane helps, a lot. But ask any pilot who's flown a red-eye after three early reports – the machine may be modern and alert, but the human brain isn't always keeping pace.

Fatigue is invisible until it isn't. It shows itself not in yawns but in missed callouts, skipped checklist steps, or a decision that lands a second too late. It trims the mental bandwidth you need when things get busy. And it does so quietly.

Visible in Flair 501, it's not always incompetence. It was the physiological reality of human brains running below capacity, an echo of what decades of research have warned us about: fatigue erodes attention, decision-making, and vigilance often before we notice it ourselves (Caldwell and Caldwell, 2016; Kanki et al., 2019; Van Dongen et al., 2003).

This chapter looks at attention, fatigue, and sleep not as abstract science but as the daily duty walk of aviation professionals. We'll look at how attention really works, how fatigue chips away at it, how sleep builds it back, and which practical strategies actually make a difference after four sectors. You don't need to be superhuman – you need to know how your mind behaves when it's tired – and learn how to work with it, not against it (Flin et al., 2008; Kanki et al., 2019).

THE SCIENCE OF ATTENTION

In aviation, few resources are as precious and as fragile as attention. It is the invisible currency that powers every checklist, every callout, every line of situational awareness. Unlike fuel or flight time, it can't be measured on a gauge, yet it dictates how safely a crew or controller performs. Attention decides what we notice, what we ignore, and how long we can keep focus before the mind drifts or narrows too much. Understanding how it works (and fails) is essential for managing fatigue, workload, and the countless micro-decisions that fill a duty day.

Types of Attention

Think of attention as a spotlight. It narrows in on what matters from the blur of cockpit noise, air traffic control chatter, and instrument cues. In aviation we lean on three modes:

- **Selective attention** – tuning in to one channel, like copying a complex clearance while cabin conversation hums in the background.

- **Divided attention** – juggling tasks, such as hand-flying while configuring and responding to ATC.
- **Sustained attention** – holding focus for long periods, whether crossing the Atlantic at night or managing a multi-sector day (Wickens and McCarley, 2019).

These modes transition and flow with workload, emotion, and fatigue. The spotlight brightens and dims.

Cognitive Limits under Pressure and Aviation-Specific Attention Failures

Working memory is not bottomless. Classic psychology puts the limit in a small range, which shrinks further under stress, time pressure, or sleep loss (Lim and Dinges, 2010; Miller, 1956). In diversions or abnormals, even seasoned crews can find the edges of their mental envelope fast.

When attention slips in the cockpit, it reflects underlying cognitive limits and typically follows familiar patterns. These may include attention being channelized, fixation and related errors, or the startle effect – different ways in which attention can become constrained under pressure. Such limits may be triggered by unexpected events, like a Traffic Collision Avoidance (TCAS) alert, that freezes action for a few critical seconds (Wickens and McCarley, 2019).

Most of these mistakes don't happen because crews lack knowledge. They happen because the brain's attention system hits its limits, especially when fatigue and stress are already pushing it to the edge (Dismukes et al., 2017; Flin et al., 2008). Those possible patterns are explained further in this chapter.

The Systemic Safety Net

Standard Operating Procedures (SOPs), checklists, and CRM exist partly to protect us from these limits. They act as external scaffolding for the moments when attention flickers. Yet even the best-designed safety nets rely on human vigilance to stay intact. When alertness fades, callouts grow quieter, cross-checks slow, and automation monitoring becomes passive.

That's why modern training increasingly treats **attention as a skill** – something to be practiced, observed, and coached, not assumed. Whether through structured scan patterns, workload distribution, or CRM briefings that include mental state awareness, the goal is to build *meta-attention*: the ability to notice when focus itself is slipping. In that sense, the science of attention isn't theoretical; it's the foundation of resilience. The chapters ahead will return to this idea, how attention interacts with fatigue, workload, and startle, and how protecting it has become one of the most modern forms of safety training (ICAO, 2020; Kanki et al., 2019).

ATTENTION LAPSES AND STARTLE EFFECTS

When Focus Fails: Understanding Attention Lapses

Attention is not an on-off switch. It drifts and fluctuates from moment to moment, shaped by workload, fatigue, circadian timing, and what's happening around us.

Even the most experienced crews are not immune. An attention lapse might last just a few seconds – but in a cockpit, those seconds matter.

Neuroergonomic research has shown that attention behaves more like a dimmer than a light switch, varying in depth and breadth (Zander and Kothe, 2011). During low workload periods like cruise, minds often wander. That "daydream drift" may feel harmless, but when an unexpected cue arrives, error rates rise sharply (Smallwood and Schooler, 2015). At the other end of the spectrum, when crews are overloaded or fatigued, they often slip into **automaticity** – running through familiar tasks on autopilot while conscious monitoring fades. This autopilot mode can create blind spots in situational awareness, especially when fatigue is layered on top (Caldwell, 2005; Van Dongen and Dinges, 2005)

Channelized Attention and Fixation Errors

Under pressure or fatigue, the brain's survival instinct is to narrow the beam of attention. In aviation we call this **channelized attention –** or the '**tunnel effect**' – where attention becomes narrowly focused on one aspect of a situation, reducing the integration of instrument data, external cues, and crew inputs (Wickens and McCarley, 2019). Crews may become absorbed in re-entering an Flight Management System (FMS) clearance and miss altitude changes, or stay glued to a non-critical system while a higher-priority task slips away.

Closely related is the **fixation error –** chasing the wrong solution with total focus, often under time pressure, while the real problem goes unnoticed (Dekker, 2017). Neither of these are about laziness or poor training; they are about cognitive overload. Working memory can only hold so much, and under stress the brain clings to what feels controllable.

A powerful example of channelised attention occurred with **Flydubai Flight 981** in 2016. After holding for nearly two hours in severe weather, the crew initiated a night-time go-around in turbulence while experiencing significant fatigue. During this high-workload manoeuvre, the crew became narrowly focused on perceived aircraft behavior, interpreting the situation as an excessive climb. In reality, sustained control inputs progressively drove the aircraft into a steep dive.

The investigation did not identify a technical failure, nor did it conclusively attribute the accident to a specific sensory illusion. Instead, investigators highlighted cognitive tunnelling (**the tunnel effect**), loss of situational awareness, degraded monitoring, and ineffective crew intervention under extreme stress. The accident demonstrated how, in high-pressure situations, attention can collapse onto a single perceived problem while critical information is no longer integrated, with catastrophic consequences (IAC, 2019).

Modern CRM increasingly teaches crews to notice these traps in real time. Metacognitive tools – asking *Where is my focus right now?* or practicing a structured scan reset – help to break tunnel vision before it snowballs (Flin et al., 2008).

The Startle Effect: Surprised into Silence

Startle is a primal response. A sudden, unexpected stimulus triggers involuntary muscle contractions, narrowed vision, a spike in heart rate, and (most critically) a pause in effective action (FAA, 2022). That pause may last only a few seconds, but in the cockpit, a few seconds can feel like an eternity.

In real-world operations, this moment rarely occurs in isolation. High-consequence events in aviation often develop within operational environments that already carry risk, shaped by a sequence of decisions made before the triggering event. These decisions are frequently reasonable when taken individually, yet when combined with sudden stress, startle, and limited time, they can define the situation crews find themselves managing only seconds later.

One example of how these dynamics interact is illustrated in the following case.

Ural Airlines Flight 178, an Airbus A321, departed Moscow Zhukovsky on 15 August 2019 with 226 passengers and crew on board. During the take-off phase, birds were present in the departure area. The take-off was continued, and seconds after lift-off the aircraft encountered a flock of birds, resulting in a near-simultaneous loss of thrust in both engines at low altitude. (IAC, 2022)

The event unfolded extremely rapidly. The bird strike was accompanied by loud noise, vibration, and multiple cockpit alerts, creating a high level of sensory and cognitive load at a critical phase of flight. According to the investigation, the crew was exposed to unexpected and acute psychological stress in the initial seconds following the engine failures. This may have affected attention, delayed procedural recall, and narrowed cognitive focus at a moment when time and altitude margins were already minimal.

In the immediate aftermath of the failure, the crew attempted to stabilise the situation under severe time pressure. The investigation noted that thrust recovery was limited and that the aircraft remained in a high-drag configuration, including the landing gear remaining extended. Within the compressed timeframe available, actions were taken sequentially as the crew attempted to manage the rapidly deteriorating energy state of the aircraft. (IAC, 2022)

The Interstate Aviation Committee (IAC; Russian acronym: MAK) report highlighted that this sequence of early actions progressively reduced the options available later in the event. It also pointed to broader contextual factors, including individual tendencies, aspects of training and operational practices, which shaped how the situation was perceived and managed in real time. The aircraft ultimately came to rest in a field near the airport, thankfully with no loss of life.

Presented here, this case illustrates how startle, extreme stress, and prior operational, cultural and individual context can interact to shape decision-making in the first seconds of an unexpected event. It is used to show how crews can become embedded in tightly constrained situations under pressure, rather than as an evaluation of technical or non-technical skills or judgement (IAC, 2022).

Simulator studies confirm that even highly experienced pilots show delayed execution when startled, particularly if the event is ambiguous or not covered by memory items ((Martin et al., 2016). This delay is not simply "being surprised" – it is a form of cognitive shock that interrupts the retrieval of trained procedures.

The industry response has been to move beyond pure technical drills. "Startle inoculation" or "Reset Method" training now includes controlled breathing, emotional self-checks, and reframing strategies – methods adapted from military and emergency medicine (Vlaskamp et al., 2024). European Union Aviation Safety Agency (EASA) and International Civil Aviation Organization (ICAO) guidance now explicitly recommends including startle-and-surprise events in training programs, not just standard failures but realistic, unexpected disruptions (EASA, 2017; ICAO, 2014).

Breaking the Loop: Training and Team Strategies

The encouraging message is that attention management can be trained. More training programs now treat it as a core skill, on par with technical flying proficiency. Among the most promising approaches are:

- **Startle inoculation with emotion regulation** – teaching crews to use breathing control and brief visualisation anchors to regain composure after a sudden event (Vlaskamp et al., 2024).
- **Metacognitive checks** – encouraging pilots to pause and ask, *Am I focused on the right thing?* during workload peaks, a technique shown to help manage cognitive tunnelling (Landman et al., 2020).
- **Wearable-based fatigue tracking** – experimental systems now use eye-tracking, reaction-time tasks, and heart-rate variability (HRV) to detect early signs of declining alertness (Adão Martins et al., 2021; McCauley et al., 2024).
- **Attention as a competency in Competency-Based Training and Assessment (CBTA)** – recent frameworks and NASA research emphasize deliberate scanning, fixation management, and focus regulation as observable skills that can be trained and assessed just like manual flying (Casner and Schooler, 2020).

The shift is subtle but important. Safety is no longer seen only as compliance with SOPs – it is also about building **cognitive readiness** (Casner and Schooler, 2020). The modern cockpit demands not just procedures, but resilience: the ability to notice when your own attention is slipping, and to recover before it matters.

UNDERSTANDING THE SPECTRUM OF FATIGUE

Fatigue is not just about feeling tired. It is the **progressive decline in alertness, judgment, and decision-making** that often goes unnoticed until safety margins thin. For crews working in dynamic and unforgiving environments, understanding fatigue as both a physiological and operational reality is essential.

Researchers typically describe fatigue in three broad categories:

- **Acute fatigue** – short-term and reversible tiredness caused by recent lack of sleep, heavy workload, or emotionally demanding events. A long duty day, a disrupted night in a noisy layover hotel, or extended troubleshooting of an abnormal can all trigger acute fatigue.
- **Chronic fatigue** – exhaustion that builds over weeks or months when recovery is inadequate or stress never truly lifts. It often carries physical symptoms, low mood, and cognitive fog. Chronic fatigue is particularly insidious in aviation because crews can normalize it; "always being tired" becomes their baseline.
- **Cumulative fatigue** – the layered effect of multiple insufficient recovery periods. Think of it as a **sleep credit card**; the more you defer repayment, the higher the interest. A string of early-morning reports, reserve shifts, and back-to-back long-hauls can quietly build this debt until performance is markedly impaired (Hereford, 2024).

Each form affects cognitive and physical functioning in slightly different ways. Acute fatigue slows reaction time and reduces vigilance. Chronic fatigue undermines motivation, stress tolerance, and judgment. Cumulative fatigue does both, and then some.

Circadian Rhythm: The Body's Internal Flight Plan

One of the strongest drivers of fatigue in aviation is the **circadian rhythm** – the body's internal 24-hour clock that regulates sleep-wake cycles, hormone release, body temperature, and alertness. Our physiology is designed for wakefulness during daylight and recovery during darkness (Graeber, 1988). Aviation often flips that expectation on its head.

Performance is particularly vulnerable during the **Window of Circadian Low (WOCL)**, usually between 0200 and 0600 local time. Core temperature dips, melatonin rises, and alertness plummets. Experimental studies have shown that reaction-time and vigilance impairments during the WOCL can approach those observed at low levels of **alcohol intoxication**, highlighting the operational significance of circadian misalignment (Arsintescu et al., 2019).

Repeated circadian disruption – whether from time-zone crossings or rotating shits and rosters – creates what is casually called *jet lag*, but in truth it is a **neurochemical disturbance**. Melatonin cycles misalign, REM sleep (rapid eye movement) becomes irregular, and the body struggles to recalibrate. The result is not only short-term fatigue but also a long-term strain on health and performance (Tsai and Lee-Chiong, 2013).

The Slippery Slope of Sleep Debt

Sleep is **the primary fuel of cognition**, and like fuel, shortages accumulate. In aviation, however, operational reality often conflicts with biological necessity. Missing even a few hours of rest does not only harm the next duty, it creates a deficit that rolls forward, quietly eroding alertness across multiple days (Caldwell, 2005).

If a pilot sleeps five hours instead of eight, those missing hours compound like interest. Over several nights, vigilance, situational awareness, and working memory deteriorate. One "catch-up nap" cannot fully reverse the decline. The same applies to air traffic controllers, whose overnight shifts and irregular rest windows quietly degrade reaction time and vigilance. Laboratory studies have repeatedly shown that moderate sleep restriction over consecutive nights produces significant performance impairments, even though participants report feeling "fine" (Van Dongen et al., 2003),

Van Dongen et al. (2006) demonstrated this clearly in a controlled performance studies: individuals subjected to chronic sleep restriction showed marked deterioration in objective task performance while remaining **confident in their self-assessments**. This disconnect between perceived and actual fatigue is among the most hazardous aspects of fatigue in aviation. It explains why external safeguards such as checklists, CRM, and peer observation are essential – we are often unable to detect our own performance degradation until it has already begun to affect safety-critical tasks.

SLEEP PHYSIOLOGY AND AVIATION REALITIES

What Sleep Really Does: The Brain on Rest

Sleep is not downtime. It's an active process where the brain and body repair, reorganize, and reset. For pilots and cabin crew, this isn't just about comfort – it's about being able to think clearly, manage stress, and make safe decisions at altitude.

Three cognitive processes depend heavily on sleep:

- **Memory consolidation** – During slow-wave and REM sleep, the brain integrates what we've learned. For aviators, that means better recall of checklists, systems knowledge, or decision strategies (Diekelmann and Born, 2010).
- **Emotional regulation** – Sleep keeps us steady. Without it, irritability rises, tolerance drops, and communication in the cockpit, cabin or office suffers (Walker, 2017).
- **Vigilance and situational awareness** – Fatigue erodes the ability to sustain attention and recognize patterns. That's when micro-sleeps, missed calls, or delayed responses show up.

Executive functions like planning, prioritizing, and risk assessment are also sharpened by restorative sleep. In aviation, these aren't luxuries – they're daily requirements.

Sleep Architecture: The Hidden Blueprint

In formal sleep science, sleep is divided into non-rapid eye movement (Non-REM or NREM) sleep, comprising stages N1, N2, and N3, and rapid eye movement (REM)

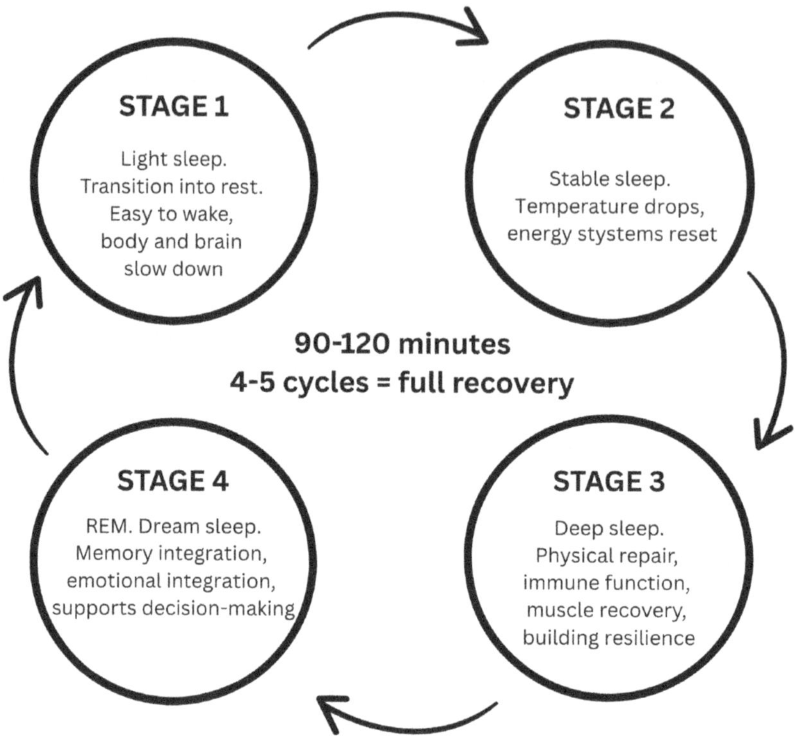

FIGURE 5.1 The four stages of a sleep cycle and their role in recovery.

sleep (Diekelmann and Born, 2010; Walker, 2017). To simplify terminology for operational and training contexts, this chapter refers to these stages using the more intuitive labels *Stage 1* (N1), *Stage 2* (N2), *Stage 3* (N3), and *Stage 4* (REM), as reflected in Figure 5.1.

Sleep runs in cycles of about 90–120 minutes, each with distinct stages (Diekelmann and Born, 2010; Walker, 2017):

- **Stage 1 (light sleep)** – The drifting-off phase, easy to wake from.
- **Stage 2** – Heart rate slows, body temperature drops, and both mind and body start recovering.
- **Stage 3 (slow-wave/deep sleep)** – Critical for physical repair, immune strength, and resilience.
- **Stage 4 – REM sleep** – Linked to vivid dreaming, emotional balance, and memory integration.

To wake up sharp, the brain needs not just enough hours but enough **complete cycles** (around 4–5). Even with "enough" total time, disruptions – hotel noise, mid-cycle wake-ups, or mismatched shifts – can strip away the deeper stages, leaving performance compromise.

Aviation Disruptors: The War on Sleep Quality

Even when the roster allows time for rest, sleep quality is often under threat. Common disruptors include:

- **Hotel environments** – Light creeping through curtains, hallway noise, thin walls, or poor bedding all fragment rest.
- **Unnatural sleep windows** – Daytime sleep before or after night flights often clashes with circadian rhythms. Even blackout blinds and earplugs rarely fully compensate.
- **Technology and screens** – Many crew, especially younger ones, unwind with Netflix, social media, or e-readers. It feels relaxing, but the blue light delays melatonin release, making it harder to fall asleep and cutting into REM quality (Cajochen et al., 2011).

For younger-generation crew in particular, "doomscrolling" before bed has become a hidden but powerful fatigue trigger. Studies show that even 30–60 minutes of evening screen exposure can delay sleep onset, suppress melatonin secretion, reduce REM sleep, and impair next-day alertness and working memory (Cajochen et al., 2011; Chang et al., 2015; Exelmans and Van den Bulck, 2016). What feels like harmless winding down often steals the very recovery their bodies need most, creating a subtle but cumulative fatigue trap that shows up the next day as slower reactions, dulled focus, or irritability.

In one controlled trial, people who read on an iPad before bed for several nights took longer to fall asleep, felt less sleepy in the evening, and were less alert in the morning compared to those who read printed books (Chang et al., 2015). In aviation, that difference matters.

Sleep Inertia and Sleep Hygiene – Aviation-Adapted Practices

Sleep inertia is the grogginess after waking – that sluggish, disoriented period when the brain hasn't yet caught up with the body. It usually lasts 10–30 minutes but can stretch longer when waking from deep sleep or during the circadian low (Walker, 2017). For aviation professionals, this lag carries real operational risk. It can strike when a standby callout jolts someone awake at 0300, when a split-duty nap ends just before reporting, or when a short hotel rest period ends mid-cycle. Physically awake doesn't always mean mentally ready, and recognizing this gap is a vital part of safe fatigue management.

However, conventional sleep advice only goes so far in aviation. Crews need tailored strategies, such as: using blackout curtains, masks, or white noise to create dark, quiet sleep spaces, avoiding caffeine for at least 6–8 hours before intended rest, limiting device use before bed or using blue-light filters/glasses when screens are unavoidable or developing wind-down rituals like stretching, breathing exercises, warm showers, or reading printed material (Caldwell et al., 2018)

Crucially, sleep hygiene is not only an individual task. It reflects airline safety culture. If rosters don't allow meaningful recovery, no amount of blackout curtains will compensate. A mature culture doesn't just meet regulatory minimums – it actively protects crew rest opportunities.

MONITORING AND MITIGATING FATIGUE

Fatigue management is no longer a single tool – it's a system of prediction, monitoring, and culture. From biomathematical models to wearables and leadership transparency, the goal is the same: recognizing when human alertness is fading and acting before performance follows.

Fatigue Risk Management Systems: The Safety Net, Evolving

Modern FRMS integrate predictive models, self-reporting, and physiological data. Tools like **SAFTE** (Sleep, Activity, Fatigue and Task Effectiveness) and **FAST** (Fatigue Avoidance Scheduling Tool) estimate alertness from sleep-wake history and circadian rhythm, highlighting when rosters may be *legal but unsafe* (Hursh et al., 2004).

Fatigue-reporting apps now let flight and cabin crew log fatigue anonymously, offering real-time insight for roster planning. Some airlines cross-reference reports with wearable data: sleep efficiency, HRV, and recovery trends – to capture fatigue patterns objectively (Adão Martins et al., 2021; Williamson et al., 2011).

Importantly, FRMS should extend beyond flight crew. Controllers, engineers, dispatchers, and ground staff face similar fatigue triggers but often lack formal reporting channels. A system-wide approach ensures consistent defenses across all operational domains.

The ICAO (2020) *Doc 9966* update emphasizes a key point: **the feedback loop must close**. When crews see how their reports shape real decisions, transparency strengthens the trust on which fatigue management depends.

THE GAP BETWEEN REGULATION AND REALITY

Even with frameworks such as **EASA Subpart FTL** and **FAA FAR 117** (EASA, 2014; FAA, 2014), fatigue remains a lived hazard. Rosters can meet limits yet still cause cumulative sleep loss – through early starts, night turns, or split duties. Younger crews sometimes hesitate to report tiredness, fearing it signals weakness, while operational or staffing pressures can normalize "legal but unsafe" schedules.

Objective testing exposes how misleading self-perception can be. Trampitsch-Vink et al. (2023) found that air-traffic controllers rated themselves alert on night shifts even as attention tests showed clear decline. Such mismatches between confidence and capacity are dangerous in aviation, where fatigue quietly narrows awareness and decision speed.

The **EASA Annual Safety Review (2025)** recognizes that psychological safety and trust in FRMS (not just regulatory compliance) are now key metrics of safety culture.

Practical Strategies in Action

Fatigue risk systems only work when individual behaviors align with them. Crews who manage fatigue effectively apply small, repeatable habits:

- **Readiness checks** before duty – "Did I sleep enough? Am I alert?"
- **Sleep banking** – extending rest before demanding rosters.
- **Energy budgeting** – using quiet phases for micro-recovery instead of distraction (Arnal et al., 2015).

They also talk about it. Simple honesty – *Three early starts, I'll lean on you today* – turns fatigue from private discomfort to shared situational awareness. Open cultures, where even junior members can raise fatigue concerns, show better cross-checking and error recovery (Williamson et al., 2011).

Digital-native crews increasingly rely on data-driven support: readiness scores, Competency-Based Training (CBT), sleep-coaching, and Evidence-Based Training (EBT) fatigue dashboards. These tools succeed only when **confidentiality** is assured; if data feel punitive, engagement collapses.

Wearables, Biometrics, and Smart Monitoring

The next frontier in fatigue mitigation is physiological and cognitive tracking. Non-invasive wearables now record sleep cycles, HRV, blinking, or micro-movements that reveal declining alertness long before it's felt. Properly used, they promote self-awareness, not surveillance.

Such systems benefit not only pilots but also air traffic controllers and maintenance staff whose shifts disrupt circadian rhythm. Research shows fatigue stems as much from monotony as from workload – quiet cruise or long night shifts can erode vigilance even with adequate sleep (Williamson et al., 2011).

The technology exists; the cultural readiness is catching up. When privacy is respected and intent is collaborative, wearable monitoring becomes a tool of empowerment – helping crews recognize and manage fatigue before safety margins fade.

Multi-layered Approach for Aviation Ecosystem

Recent reviews confirm that biomathematical models alone cannot predict human performance. The most reliable defense is a **multi-layered approach** – combining predictive data, wearables, and honest self-reporting (Kakhi et al., 2024; McCauley et al., 2024)

Before fatigue can be managed, it must be understood from every angle.

No single dataset – whether predictive models, biometrics, or self-reports – captures the full picture of human alertness. Effective management integrates insights from multiple sources: the regulatory framework that defines limits, the systems that model risk, the technology that senses physiological change, the organizational data that shape rosters, and the culture that encourages honest reporting.

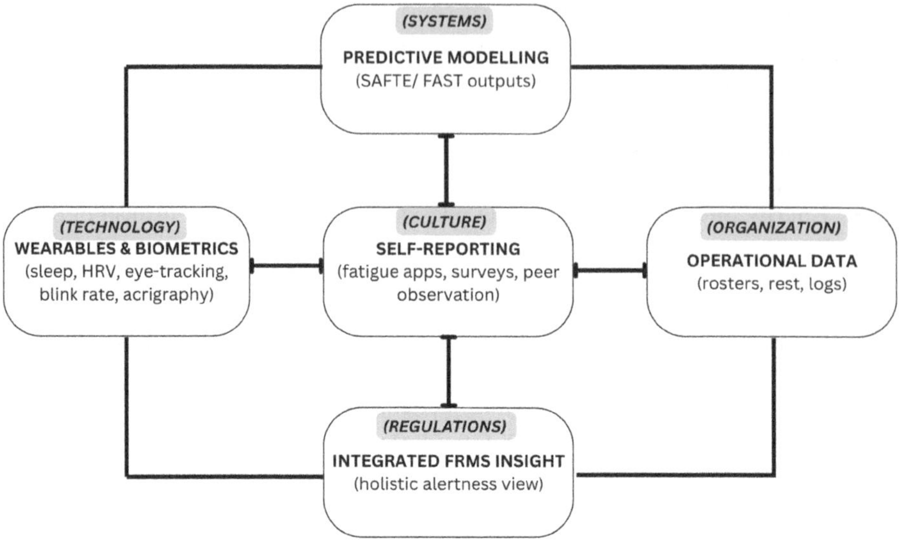

FIGURE 5.2 Multi-dimensional fatigue management ecosystem.

Together, these layers create a **living ecosystem of fatigue awareness** rather than a static rule set.

In the end, fatigue management is not about eliminating tiredness but about understanding it – turning awareness into protection. Fatigue will always be part of human performance, but how we face it defines our safety. When data, empathy, and leadership work together, aviation moves beyond managing fatigue – it learns to outsmart it (Figure 5.2).

TAKEAWAYS

For Pilots and Cabin Crew

- Name fatigue early – in briefings or mid-flight. It's not weakness, it's awareness. Set the tone by saying it out loud before performance is compromised.
- Bank your sleep before demanding rosters. Use 2–3 nights of extended sleep before early starts or long-hauls to protect your performance.
- Try a readiness app, a wearable, or even a simple sleep journal to understand your own patterns and weak spots.

Reflective Question*: When was the last time I missed a small cue or forgot a step – and could it have been fatigue? What were the early warning signs I didn't notice?*

For Instructors, Examiners, and Leaders

- Let your trainees or personnel see you acknowledge tiredness in a professional, proactive way – it makes it normal and safe for them to do the same.
- Embed startle realism. Surprise scenarios should reflect today's realities – not just repeating the same scenarios, but think about cognitive overload and ambiguous failures.
- Go beyond procedures – discuss how fixation, scanning habits, and "zoom-out" thinking can make or break situational awareness.

Reflective Question: *Do I treat fatigue and attention breakdowns as teachable skills – or still view them as personal shortcomings?*

For Operational Teams

- Meeting duty limits is the baseline, not the goal. Build rosters using fatigue risk forecasts, crew feedback, and actual sleep data where possible, because legal does not always mean safe.
- Encourage the use of fatigue-monitoring or wellness wearables as voluntary performance aids, not surveillance instruments. Framing them as support tools fosters trust and uptake.
- Reinforce through policy, training, and daily leadership that reporting fatigue is a sign of professionalism, not weakness. Most incidents prevented by fatigue reports never make headlines – and that's exactly the point.

Reflective Question: *What message does our system send: that fatigue is expected and acknowledged, or denied and punished?*

REFERENCES

Adão Martins, N. R., Annaheim, S., Spengler, C. M. and Rossi, R. M. (2021) 'Fatigue monitoring through wearables: A state-of-the-art review', *Frontiers in Physiology*, 12, p. 790292. https://doi.org/10.3389/fphys.2021.790292

Arnal, P. J., Sauvet, F., Léger, D., Van Beers, P., Bayon, V., Bougard, C., Rabat, A., Millet, G. Y. and Chennaoui, M. (2015) 'Benefits of sleep extension on sustained attention and sleep pressure before and during total sleep deprivation and recovery', *Sleep*, 38(12), pp. 1935–1943. https://doi.org/10.5665/sleep.5244

Arsintescu, L., Kato, K. H., Hilditch, C. J., Gregory, K. B. and Flynn-Evans, E. (2019) 'Collecting sleep, circadian, fatigue, and performance data in complex operational environments', *Journal of Visualized Experiments*, (150), e59851, https://doi.org/10.3791/59851

Cajochen, C., Frey, S., Anders, D., Späti, J., Bues, M., Pross, A. and Stefani, O. (2011) 'Evening exposure to an LED-backlit screen affects circadian physiology and cognitive performance', *Journal of Applied Physiology*, 110(5), pp. 1432–1438. https://doi.org/10.1152/japplphysiol.00165.2011

Caldwell, J. A. (2005) 'Fatigue in aviation', *Travel Medicine and Infectious Disease*, 3(2), pp. 85–96. https://doi.org/10.1016/j.tmaid.2004.07.008

Caldwell, J. A. and Caldwell, J. L. (2016) *Fatigue in Aviation: A Guide to Staying Awake at the Stick*. Abingdon: Routledge. https://doi.org/10.4324/9781315582030

Caldwell, J. A., Caldwell, J. L., Thompson, L. A. and Lieberman, H. R. (2018) 'Fatigue and its management in the workplace', *Neuroscience & Biobehavioral Reviews*, 96, pp. 272–289. https://doi.org/10.1016/j.neubiorev.2018.10.024

Casner, S. M. and Schooler, J. W. (2020) *Training for Attention Management in Flight Operations (NASA/TM-2020-5003444)*. Moffett Field, CA: NASA Ames Research Center. Available at: https://ntrs.nasa.gov/api/citations/20205003444/downloads/NASA-TM-20205003444.pdf

Chang, A.-M., Aeschbach, D., Duffy, J. F. and Czeisler, C. A. (2015) 'Evening use of light-emitting eReaders negatively affects sleep, circadian timing and next-morning alertness', *PNAS*, 112(4), pp. 1232–1237. https://doi.org/10.1073/pnas.1418490112

Dekker, S. (2017) *The Field Guide to Understanding Human Error*. 3rd ed. Boca Raton, FL: CRC Press. https://doi.org/10.1201/9781315239675

Diekelmann, S. and Born, J. (2010) 'The memory function of sleep', *Nature Reviews Neuroscience*, 11(2), pp. 114–126. https://doi.org/10.1038/nrn2762

Dismukes, R. K., Berman, B. A. and Loukopoulos, L. (2017) *The Limits of Expertise*. Abingdon: Routledge. http://dx.doi.org/10.4324/9781315238654

European Union Aviation Safety Agency (EASA) (2014) *Commission Regulation (EU) No 83/2014 of 29 January 2014 Amending Regulation (EU) No 965/2012 as Regards Flight and Duty Time Limitations and Rest Requirements (Subpart FTL)*. Cologne: European Union Aviation Safety Agency. Available at: https://eur-lex.europa.eu/LexUriServ/LexUriServ.do?uri=OJ:L:2014:028:0017:0029:EN:PDF

European Union Aviation Safety Agency (EASA) (2017) *EASA Research Project: The Effect of Startle and Surprise on Flight Crew Performance (EASA.2013.OP.06)*. Cologne: European Union Aviation Safety Agency. Available at: https://www.easa.europa.eu/sites/default/files/dfu/EASA_Research_Startle_Effect_Managements_Final_Report.pdf

European Union Aviation Safety Agency (EASA) (2025) *Annual Safety Review 2025*. Cologne: European Union Aviation Safety Agency. Available at: https://www.easa.europa.eu/en/document-library/general-publications/annual-safety-review-2025

Exelmans, L. and Van den Bulck, J. (2016) 'Bedtime mobile phone use and sleep in adults', *Social Science & Medicine*, 148, pp. 93–101. https://doi.org/10.1016/j.socscimed.2015.11.037

Federal Aviation Administration (FAA) (2014) *Flightcrew Member Duty and Rest Requirements (14 CFR Part 117)*. Washington, DC: Federal Aviation Administration. Available at: https://www.ecfr.gov/current/title-14/chapter-I/subchapter-G/part-117

Federal Aviation Administration (FAA) (2021) *Startle Response in Aviation*. Washington, DC: Federal Aviation Administration. Available at: https://www.faa.gov/sites/faa.gov/files/2022-01/Startle%20Response.pdf

Flin, R., O'Connor, P. and Crichton, M. (2008) *Safety at the Sharp End: A Guide to Non-Technical Skills*. Boca Raton, FL: CRC Press. https://doi.org/10.1201/9781315607467

Graeber, R. C. (1988) 'Aircrew fatigue and circadian rhythmicity', in E. L. Wiener, and D. C. Nagel (Eds.), *Human Factors in Aviation*. San Diego, CA: Academic Press, pp. 305–344. https://doi.org/10.1016/B978-0-08-057090-7.50016-3

Hereford, J. M. (2024) 'Consequences of sleep deprivation', in Hereford, J. (ed.) *Sleep and Rehabilitation*. Abingdon: Routledge, pp. 103–112. https://doi.org/10.4324/9781003526438-14

Hursh, S. R., Redmond, D. P., Johnson, M. L., Thorne, D. R., Belenky, G., Balkin, T. J., Storm, W. F., Miller, J. C. and Eddy, D. R. (2004) 'Fatigue models for applied research in warfighting', *Aviation, Space, and Environmental Medicine*, 75(3 Suppl), pp. A44–A60.

International Civil Aviation Organization ICAO. (2014) *Manual on Aeroplane Upset Prevention and Recovery Training (Doc 10011)*. Montreal: International Civil Aviation Organization.

International Civil Aviation Organization (ICAO) (2020) *Manual for the Oversight of Fatigue Management Approaches*, Doc 9966, Second edition, version 2 (revised). Montreal: ICAO. Available at: https://elibrary.icao.int/product/234625 / https://store.icao.int/en/manual-for-the-oversight-of-fatigue-management-approaches-doc-9966

Interstate Aviation Committee (IAC). (2019) *Final Report on the Accident Involving Boeing 737-800 A6-FDN Operated by FlyDubai at Rostov-on-Don Airport (Russia) on 19 March 2016*. Moscow: IAC. Available at: https://mak-iac.org/upload/iblock/1e8/report_a6-fdn_eng.pdf

Interstate Aviation Committee (IAC) (2022) *Final Report on the Results of the Aviation Incident Investigation Involving Airbus A321-211 (VQ-BOZ) Operated by Ural Airlines Near Zhukovsky Airport, Russia, on 15 August 2019*. Moscow: Aircraft Accident Investigation Commission, Interstate Aviation Committee. Available at: https://centrostudi-stasa.com/wp-content/uploads/2024/10/MAK.-2022.-FINAL-REPORT-ON-THE-RESULTS-OF-THE-AVIATION-INCIDENT-INVESTIGATION.-Ural-Airlines-178-VQ-BOZ-15-August-2019.pdf

Kakhi, K., Jagatheesaperumal, S. K., Khosravi, A., Alizadehsani, R. and Acharya, U. R. (2024) 'Fatigue monitoring using wearables and AI: Trends, challenges, and future opportunities', *Expert Systems with Applications*, 247, pp. 123 785. https://doi.org/10.1016/j.compbiomed.2025.110461

Kanki, B. G., Helmreich, R. L. and Anca, J. (2019) *Crew Resource Management*. 3rd ed. Cambridge, MA: Academic Press.

Landman, A., van Rooij, J. and Vlaskamp, D. (2020) 'The effectiveness of a mnemonic-type startle and surprise procedure', *Aviation Psychology and Applied Human Factors*, 10(1), pp. 3–12. https://doi.org/10.1080/24721840.2020.1763798

Lim, J. and Dinges, D. F. (2010) 'A meta-analysis of the impact of short-term sleep deprivation on cognitive variables', *Psychological Bulletin*, 136(3), pp. 375–389. https://doi.org/10.1037/a0018883

Martin, W., Murray, P., Bates, P. and Lee, S. Y. (2016) 'A flight simulator study of the impairment effects of startle on pilots during unexpected critical events', *Aviation Psychology and Applied Human Factors*, 6(1), pp. 24–32. https://doi.org/10.1027/2192-0923/a000092

McCauley, M. E., McCauley, P., Kalachev, L. V., Riedy, S. M., Banks, S., Ecker, A. J., Dinges, D. F. and Van Dongen, H. P. A. (2024) 'Biomathematical modeling of fatigue due to sleep inertia', *Journal of Theoretical Biology*, 590, p. 111851. https://doi.org/10.1016/j.jtbi.2024.111851

Miller, G. A. (1956) 'The magical number seven, plus or minus two: Some limits on our capacity for processing information', *Psychological Review*, 63(2), pp. 81–97. https://doi.org/10.1037/h0043158

Smallwood, J. and Schooler, J. W. (2015) 'The science of mind wandering', *Annual Review of Psychology*, 66, pp. 487–518. https://doi.org/10.1146/annurev-psych-010814-015331

Trampitsch-Vink, L. S., Riss, H., Paul, K. and Raggam, R. B. (2023) 'Fatigue in operational nightshifts', *Somnologie*, 27(2), pp. 65–72. https://doi.org/10.1007/s11818-023-00434-3

Tsai, S. C. and Lee-Chiong, T. Jr. (2013) 'Sleep disorders and fatigue', *Sleep Medicine Clinics*, 8(2), pp. 235–239. https://doi.org/10.1016/j.jsmc.2013.02.003

Van Dongen, H. P. A., Caldwell, J. A. and Caldwell, J. L. (2006) 'Investigating systematic individual differences in sleep-deprived performance on a high-fidelity flight simulator', *Behavior Research Methods*, 38, pp. 333–343. https://doi.org/10.3758/BF03192785

Van Dongen, H. P. A. and Dinges, D. F. (2005) 'Sleep, circadian rhythms, and psychomotor vigilance', *Clinics in Sports Medicine*, 24(2), pp. 237–249. https://doi.org/10.1016/j.csm.2004.12.007

Van Dongen, H. P. A., Maislin, G., Mullington, J. M. and Dinges, D. F. (2003) 'Cumulative cost of additional wakefulness', *Sleep*, 26(2), pp. 117–126. https://academic.oup.com/sleep/article/26/2/117/2709341

Vlaskamp, D., Landman, A., van Rooij, J. and Blundell, J. (2024) 'Recovery from startle and surprise: A survey of airline pilots' operational experience using a startle and surprise recovery method'. *SSRN Electronic Journal*. https://doi.org/10.2139/ssrn.5017016

Walker, M. (2017) *Why We Sleep: Unlocking the Power of Sleep and Dreams*. New York: Scribner.

Wickens, C. D. and McCarley, J. S. (2019) *Applied Attention Theory*. Abingdon: Routledge. https://doi.org/10.1201/9780429059261

Williamson, A., Lombardi, D. A., Folkard, S., Stutts, J., Courtney, T. K. and Connor, J. L. (2011) 'The link between fatigue and safety', *Accident Analysis & Prevention*, 43(2), pp. 498–515. https://doi.org/10.1016/J.AAP.2009.11.011

Zander, T. O. and Kothe, C. (2011) 'Towards passive brain-computer interfaces: Applying brain-computer interface technology to human–machine systems in general', *Journal of Neural Engineering*, 8(2), p. 025005. https://doi.org/10.1088/1741-2560/8/2/025005

6 Workload and Situational Awareness
Conscious Skies

CASE STUDY: AIR CANADA FLIGHT 759 (2017) – MOMENTS FROM DISASTER

Late in the evening of July 7, 2017, Air Canada Flight 759, an Airbus A320, was inbound to San Francisco International Airport (SFO) after a routine flight from Toronto. The weather could not have been more forgiving – clear skies, calm winds, and excellent visibility. Yet, inside the cockpit, conditions were far from optimal. The crew were battling the weight of accumulated fatigue and the mental demands of a busy arrival into one of the most complex airfields in the United States, with Runway 28L closed, creating an unusual visual environment (NTSB, 2018).

As the aircraft descended toward the Bay Area, the pilots' workload intensified. The Airbus was cleared for Runway 28R, but instead of lining up on the correct runway, the crew aligned with parallel Taxiway C. On that taxiway sat four fully loaded airliners, waiting their turn for departure (NTSB, 2018).

For several seconds, the crew's perception did not match reality. The unusual lighting cues sparked brief uncertainty – was the spacing right, were they aligned correctly? – but the situation was interpreted in line with their expectation of a runway ahead. The cockpit voice recorder later revealed hesitations and half-formed questions, but not the decisive intervention that the situation demanded (NTSB, 2018).

Inside one of the aircraft on the taxiway, another pilot instinctively switched on landing lights, an act of both self-protection and warning. Almost at the same time, the controller in the SFO tower noticed the unsettling alignment and asked Flight 759 to confirm their runway. That question cut through the fog of misperception. In an instant, realization dawned in the cockpit. The crew initiated a go-around, hauling the Airbus skyward. At its lowest point, the A320 passed within roughly 60 feet of the first aircraft on the taxiway – a margin measured in seconds and human lives (NTSB, 2018).

The National Transportation Safety Board (NTSB) later concluded that a dangerous mix of fatigue, workload saturation, reliance on visual rather than instrument cues, and unfamiliarity with recent airport changes had eroded the crew's situational awareness (SA) (NTSB, 2018). This incident stands as one

DOI: 10.1201/9781003515418-6

of aviation's closest near-misses, a chilling reminder that workload and SA are not abstract concepts but fragile threads holding safety together. When stretched too far, they fray alarmingly fast.

INTRODUCTION

Few incidents show how thin aviation safety margins can be as clearly as Air Canada Flight 759. What unfolded was not weather, failure, or rare technical quirks, but the intersection of two invisible forces that shape every operation: workload and SA. Both rise and fall quietly, influencing performance long before errors are visible.

In human factors, **workload** is not the number of tasks but how demands arrive, overlap, and compete for limited attention and working memory (FAA, 2016; Wickens, 2008). When demand is moderate and time allows, performance is stable; when demand spikes or resources shrink through fatigue or distraction, it frays. The signs are often hidden: a pilot appears calm while juggling modes and calls; a controller sounds composed while managing converging traffic; a cabin or maintenance team reorders priorities under time pressure.

SA is equally difficult to identify: a three-layered process of perception, comprehension, and projection (Endsley, 1995) – and it can thin under fatigue, distraction, expectation bias, or overconfidence just when clarity matters most. Decades of research keep returning to these constructs because they link directly to performance and safety outcomes across flight deck, cabin, tower, and hangar (Young et al., 2015).

Understanding how workload builds and how SA fades is not only for investigation. It underpins modern Competency-Based Training and Assessment (CBTA) and operational design that anticipates human limits rather than punishes them. Managed well, workload and SA act as quiet stabilizers, usually invisible, decisive when the unexpected appears.

WHAT WORKLOAD REALLY MEANS

Workload is one of those concepts everyone in aviation recognizes but few can define precisely. After a demanding approach or a congested sector, people often say, "It was too much at once." In human factors, workload is not just about how many tasks exist, but how they arrive, overlap, and interact with the limited resources of attention and working memory.

In practical terms, workload reflects the **relationship between task demand and the operator's available capacity** (Pereira da Silva, 2014). When demand is moderate and time allows, performance is stable. When demand spikes or resources shrink through fatigue or distraction, performance begins to be affected (Young et al., 2015). The difficulty is that workload can be invisible. A pilot may look calm while juggling automation modes and ATC calls; a controller may sound composed while tracking multiple aircraft; a cabin crew member may appear unhurried while prioritizing safety checks, service, and passenger issues.

Dimensions of Workload

Workload is multidimensional: **mental, physical, temporal, and emotional** (Hart, 2006; Masi et al., 2023). The mix of these dimensions, rather than their number, defines how demanding a situation feels.

1. **Mental workload** dominates most modern operations. Across roles, professionals juggle data, planning, and priorities under time pressure. Mental load is the invisible weight of thinking, often the first to trigger errors when capacity runs thin (Jafari et al., 2019).
2. **Physical workload** is less visible in the flight deck but significant elsewhere. A go-around or manual approach increases muscular effort and coordination. Cabin crew spend long hours standing, handling equipment, assisting passengers, and conducting safety tasks in tight spaces. Engineers work in constrained conditions, often in poor weather or limited light (Mulay et al., 2019; Williamson et al., 2011).
3. **Temporal workload** reflects how quickly demands arrive. Aviation compresses time – a manageable sequence can turn chaotic when calls, changes, or cabin events cluster within seconds (Hancock and Desmond, 2001).
4. **Emotional workload** is the least discussed yet deeply influential. A disruptive passenger, a sick child onboard, or an unresolved near-miss all impose strain that competes with attention. Emotional effort connects directly to resilience and long-term well-being (Masi et al., 2023).

These layers rarely act alone; they blend. A maintenance engineer troubleshooting under time pressure may face physical strain, mental load, and emotional tension all at once. Together, these overlapping demands form the practical reality of work across different roles in aviation, where physical, mental, temporal, and emotional effort constantly interact and shape performance

A complementary framework, **Multiple Resource Theory** (Wickens, 2002), helps explain why overload happens. The human brain cannot handle unlimited parallel input: visual, auditory, and motor tasks draw from partially separate but finite pools. Reading an approach plate while listening to ATC competes for the same resources; flying manually while listening is easier because it divides channels. When too many inputs draw from the same pool, interference builds, and SA starts to thin (Figure 6.1).

Workload Curve – Under and Overload

The relationship between workload and performance follows an **inverted-U pattern**, described by the **Yerkes-Dodson law** (Diamond et al., 2007; Yerkes and Dodson,

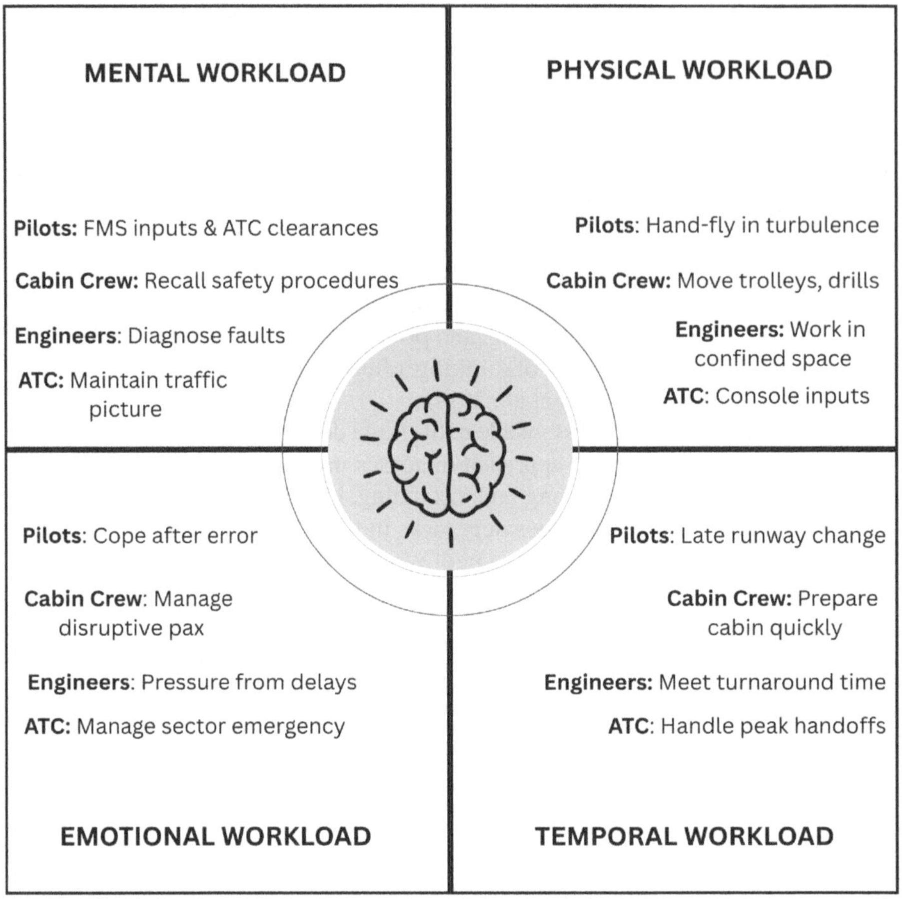

FIGURE 6.1 Dimensions of workload in aviation roles.

1908). At low workload, vigilance fades and the mind drifts; at moderate levels, focus sharpens and performance peaks. Push further, and the curve drops: attention narrows, working memory collapses, and small mistakes multiply.

This curve also expands on what was introduced earlier in the previous chapter. It explored how low stimulation and monotony degrade alertness over time; here, the emphasis shifts to how **mental load** dynamically shapes performance. Both perspectives point to the same truth: safety depends on staying in the middle zone where effort and capacity remain balanced.

Cognitive science shows why this equilibrium is so fragile. **Working memory** has very limited spare capacity; when calls, alarms, and competing tasks cluster, the instinct to work faster usually makes things worse (Dismukes et al., 2007). The safer response is to slow down and regain cognitive control.

Across aviation roles, similar strategies appear. A pilot on approach may delay making a passenger announcement until the aircraft is stable; cabin crew may postpone a non-essential service to secure the cabin first; an air-traffic controller under pressure may pause and request a repeat clearance to regulate tempo; and a maintenance engineer may jot a reminder rather than hold it in memory. Each reflects the same discipline, known as **deliberate task shedding** to keep performance within the optimal zone.

The opposite risk, **underload**, is quieter but no less hazardous. Long, uneventful stretches reduce stimulation and can cause the brain to drift, a phenomenon once described as the ***vigilance decrement*** (Mackworth, 1948; Warm et al., 2008). It feels different from fatigue – the eyes still move, but awareness fades. To counter it, operators build small moments of re-engagement: pilots summarize their position, fuel, and traffic aloud; controllers rehearse the next handover; cabin crew mentally run through diversion scenarios. Even small adjustments: standing, stretching, or consciously resetting a scan – help bring the system back toward the top of the curve.

Modern CBTA and Evidence-Based Training (EBT) approaches integrate these lessons directly. Figure 6.2 illustrates this relationship through a conceptual workload-performance curve commonly used in aviation psychology. Crews now practice both high-tempo and low-tempo situations to recognize when their performance is sliding toward either extreme (overload or underload) and to recover deliberately before awareness thins (EASA, 2023).

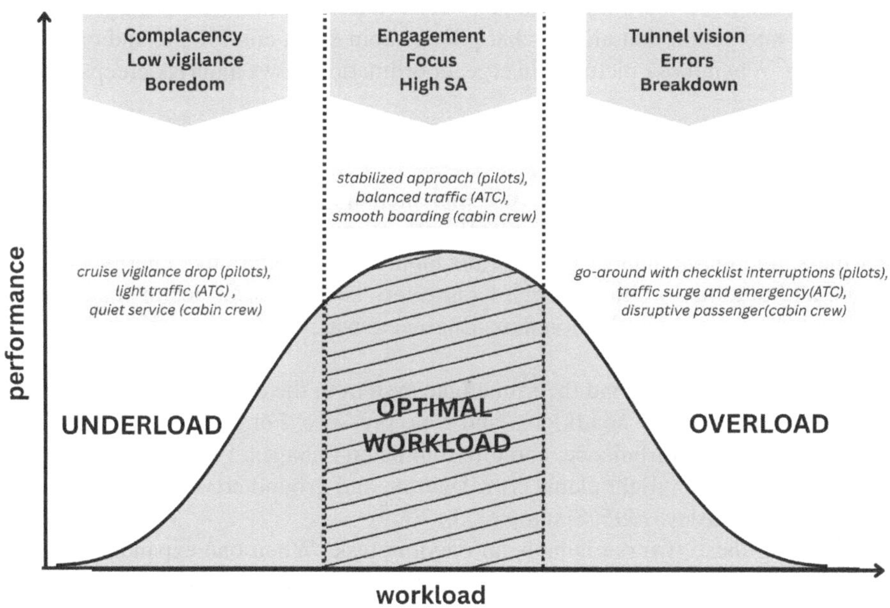

FIGURE 6.2 The workload curve in aviation context. Author's conceptual synthesis informed by workload and arousal theories.

Coping at the Edges

When workload spikes, the instinct is to speed up. The safer strategy is to slow the tempo and shed tasks deliberately. **Aviate – navigate – communicate** exists for that reason – to anchor attention on essentials and make priorities explicit.

Across aviation roles, the principle is the same:

- Flight crews delay non-essential calls or checklists until stable.
- Cabin crew defer secondary tasks when safety priorities rise.
- Controllers regulate tempo using standard phrases.
- Engineers mark visible checkpoints to avoid losing sequence after interruption.

At the other end, long calm periods invite underload, so vigilance loss and mental drift (Warm et al., 2008). Operators counter this by staying mentally active: pilots refresh position summaries, controllers rehearse conflict resolutions, cabin crew mentally review evacuation plans. The aim is not to stay busy but to stay present.

DEFINING SITUATIONAL AWARENESS

SA is a constantly shifting mental model – the internal picture that lets an operator understand what is happening, what it means, and what will come next. Endsley (1995) defined SA as a three-stage process: **perception**, **comprehension**, and **projection**. Each stage is vulnerable to overload, distraction, and bias.

In aviation, SA is both individual and collective, like the mental bridge between awareness and action that aligns what pilots, cabin crew, controllers, and engineers see and do. When those pictures diverge, coordination slows and risk creeps in.

Situational Awareness – Inside and Outside Perspective

It helps to think of SA as having two complementary domains – **inside** and **outside**.

Inside SA is system-focused: aircraft state, automation modes, checklists, cabin readiness, and maintenance status. Pilots monitor parameters like energy, configuration, and automation mode logic. Controllers keep track of sector load, equipment status, and communication flow. Engineers watch system trends and deferred items.

Outside SA looks beyond the immediate system to the operational environment: weather, traffic, runway conditions, and time pressure. For cabin crew, it includes passenger behavior, turbulence, and environmental changes. For operations control or dispatch, it means flight plans, crew legality, and ground constraints (Durso and Dattel, 2006; Endsley, 1995; Stanton et al., 2009).

Balancing these two domains is an ongoing task. When one expands, the other tends to shrink. A pilot deep in Flight Management System (FMS) re-programming (inside) can momentarily lose the outside picture of traffic or terrain. Conversely, prolonged head-up scanning in deteriorating weather can lead to mode confusion or missed checklist items. Accident and incident reports are filled with examples where one domain quietly dominated until it was too late (Endsley and Jones, 2012).

The Forward Edge – Anticipation and Shared Awareness

Strong operators are not only aware of what is happening – they stay slightly ahead of it. The third layer of SA, **projection**, keeps them anticipating the next clearance, call, or change before it occurs. This forward edge transforms awareness into foresight.

Micro-projections are rarely dramatic. They surface as short, factual cues that align the team: "*expect a late downwind; plan a long final*," "*if turbulence worsens, we'll secure early*." Such phrases synchronize mental models across roles – what Endsley and Jones (2012) call **shared projection**. In flight decks, cabins, and control rooms, they create a common rhythm of thought. When SA fades, these cues vanish, and talk turns reactive: a subtle sign that fatigue or workload is closing the gap between thought and action.

Projection relies on discipline and communication. Declaring priorities aloud – *"Holding that call," "We'll brief again at top of descent"* – structures thought and buys cognitive margin. Effective Crew Resource Management (CRM) depends on this habit: voicing what you're thinking so others can update their own models.

Fatigue, stress, and overconfidence each blunt anticipation differently. Fatigue shortens attention, stress narrows perception, and overconfidence dulls curiosity. None destroys SA outright, but they flatten its forward edge until teams react rather than anticipate.

Aviation's resilience lies in **collective vigilance**. When one person catches what another missed: a controller notices spacing drift, or a cabin colleague spots a skipped check – shared awareness is restored. That recovery is not luck but design: a system built on distributed cognition, where safety emerges from many minds tracking the same goal (Endsley and Jones, 2012).

Design and Human Limits – When the System Helps or Hinders

SA depends as much on how information is displayed as on who interprets it. Modern aircraft, ATC systems, and maintenance tools produce vast data streams; the challenge is turning them into meaning, quickly enough to act. Integrated displays showing energy, path, and mode ease cognitive load, while cluttered layouts or hidden states delay comprehension (Endsley and Jones, 2012; Sarter and Woods, 1995).

Intuitive design means crews need not rebuild the picture from memory. Clear mode annunciations, logical alert hierarchies, or trend-based maintenance data preserve SA under pressure. Poorly designed automation, by contrast, breaks the feedback loop – mode changes that go unnoticed leave operators "out of the loop," supervising systems they no longer understand (Bainbridge, 1983).

The same logic extends beyond the cockpit. Cabin panels showing combined alerts or dispatch tools emphasizing time-critical data protect SA by simplifying decision flow. Clarity of context matters more than the amount of data.

Even the best design cannot offset human limits. Fatigue, workload, and emotional strain narrow scans and slow updates (Loft et al., 2007; Wickens, 2002). Transparency and simplicity remain the strongest safeguards – whether in interfaces or organizational processes.

WHY WE LOSE SITUATIONAL AWARENESS

SA rarely collapses in a single moment. More often, it thins gradually as attention narrows, communication fades, and assumptions quietly harden. Across operational roles, recurring triggers appear: workload spikes, interruptions, unclear automation behavior, misaligned mental models, and complacency from routine. These influences overlap, eroding perception, comprehension, and projection until the picture no longer aligns with reality.

COGNITIVE AND ENVIRONMENTAL PRESSURES

The first and most visible threat is **cognitive overload**. When information and demands arrive together, working memory clogs and attention narrows to a single thread. Pilots may fixate on an FMS entry while missing traffic cues; controllers may chase one conflict while sector awareness fades. These moments are not lapses of discipline but natural limits of bandwidth (Parasuraman and Hancock, 2001; Wickens, 2002).

Environmental pressure – multiple radio calls, chimes, or cabin demands – adds to the squeeze. The safer move is to slow down and shed tasks deliberately. Declaring priorities aloud ("Aviate first, we'll brief later") preserves cognitive margin (Dismukes et al., 2007).

Interruptions worsen the effect: each pause requires time to rebuild mental context. A missed call-sign or checklist line can trace back to this lag (Bailey et al., 2001). Visible hold points in checklists or verbal markers ("We'll pick up after line four") help restore continuity. Sterile cockpit rules and quiet zones serve the same purpose – to protect fragile cognitive threads.

SYSTEM AND TEAM DISCONNECTS

Even with spare capacity, awareness can fragment between humans and systems. Automation reduces physical work but adds cognitive risk: the out-of-the-loop problem (Bainbridge, 1983). When modes switch quietly or cues are vague, operators supervise without true comprehension. The familiar "how did we get into that mode?" moment reflects design, not negligence (Sarter and Woods, 1995).

Transparent feedback – sharp mode announcements, alerts, and clear hierarchies – keeps people aligned with what automation is doing. Cluttered or ambiguous interfaces stretch comprehension until meaning lags events (Endsley and Jones, 2012).

Teams face the same risk: **misaligned mental models**. Two pilots may sound coordinated yet imagine different futures – one a short base, the other a long final. Controllers, cabin crew, and engineers encounter similar drift as situations evolve differently in their heads. These gaps fracture shared SA and slow decisions (Cooke et al., 2007; Mathieu et al., 2000).

The direct projection is a countermeasure. Short forward statements align perspectives – "Expect delay vectors, manage energy," "If the passenger deteriorates, we divert." Light, factual anchors keep everyone in the same frame.

The Drift: Complacency and Misjudged Awareness

When systems run smoothly, vigilance dulls. Reliability invites the quiet assumption that all is well. In cockpits, ATC, and engineering, this complacency drift is a normal cognitive response to routine, not moral weakness (Parasuraman et al., 1993). Mental models flatten, raw-data checks fade, and forward projection thins.

Strong operators counter with light **re-anchoring**: cross-check instruments against expectations, scan raw data, and run brief what-if questions during calm phases. These resets restore healthy tension without adding clutter.

Confidence is not the same as SA. Post-event ratings or self-assessments often reflect certainty, not understanding (Endsley, 1995). Eye-tracking, scan metrics, or brief probes in simulation can expose gaps between perception and performance, though over-measurement can itself distort attention.

For training organizations, triangulate: **compare self-ratings with observed behavior and outcomes**. If the plan deviated from reality, awareness was thin regardless of confidence. The goal is not grading but mirrors – feedback loops that reveal fading SA early, so it can be rebuilt.

These mechanisms – overload, disconnection, and drift – form the **SA Erosion Funnel** (Figure 6.3). Each layer narrows perception until the operator sees less, reacts slower, and believes more firmly in a partial picture. Recovery starts by spotting the layer: lighten load, reconnect the model, or re-anchor attention before projection disappears.

INDIVIDUAL DIFFERENCES IN HANDLING WORKLOAD AND SA

No checklist or system fully levels out how people process pressure. Even when training, equipment, and conditions are standardized, human performance varies widely. Some operators absorb a heavy tempo with ease; others reach saturation under the same load. These contrasts are not flaws but reflections of how cognitive resources, temperament, and experience combine differently in each person.

Understanding those differences matters because they shape how workload and SA rise and fall in real time. Recognizing them allows instructors, assessors, and operators to tailor strategies – protecting performance margins not by equalizing everyone, but by helping each person work with their own strengths and limits.

The Spectrum of Cognitive Differences

Line operations remind us of a simple truth: people don't carry identical cognitive toolkits. Two pilots can fly the same sector under identical conditions and experience it entirely differently – one comfortably ahead of the aircraft, the other close to saturation. Controllers, cabin leaders, and engineers all recognize the same pattern: some thrive on complexity; others need calm and structure to regain clarity.

Aviation psychology shows that four broad categories of individual difference matter most for workload and SA: cognitive capacity, work style, personality, and experience (Carretta and Ree, 2003). Together they explain why some operators appear calm and composed while others feel overloaded under identical task demands.

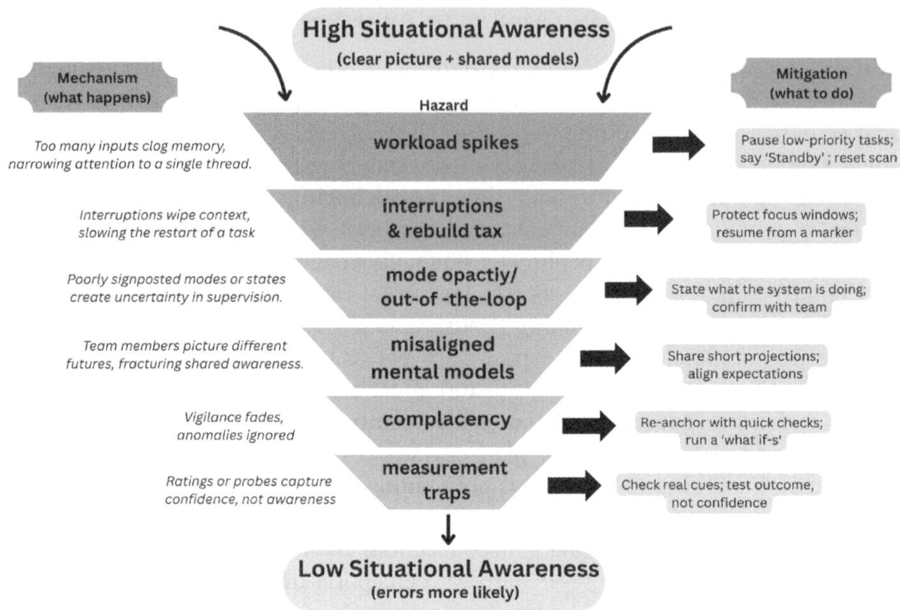

FIGURE 6.3 SA Erosion Funnel – mechanisms, hazards, and mitigations.

Cognitive capacity and work style form the mental foundation. Working memory is the brain's short-term stage for aviation: it holds clearances while programming the box, keeps traffic intentions alive while monitoring calls, and tracks passenger tasks during descent. Capacity varies widely (Engle, 2018; Kane and Engle, 2002). Those with greater capacity can juggle competing inputs without slipping into fixation, whereas limited capacity makes tunnel vision more likely (Endsley and Bolstad, 1994).

Style shapes how that capacity is used. Some operators prefer rapid task-switching (**polychronicity**); others perform best when running tasks sequentially to a clear checkpoint (Lievens et al., 2008). In simulation studies, differences in updating, inhibition, and spatial working memory predict who adapts most smoothly as workload multiplies (Morgan et al., 2013). High-polychronic individuals might handle several tasks at once until a mode change disrupts them; sequential operators may keep SA cleaner by requesting vectors or delaying setup to finish one step properly. Good instruction respects both tendencies – teaching boundaries rather than imposing a single "right" style.

Personality and Behavior under Pressure

Personality is not destiny, but it steers behavior at the margins. **Conscientiousness** and related self-regulatory traits correlate with safer performance and fewer incidents, while **neuroticism** and **impulsivity** increase risk-taking under strain (Beus et al., 2015; Christian et al., 2009; Clarke and Robertson, 2005). In everyday language, one voice says, "Let's build margin," the other, "We can make it."

In air traffic control, prevention-focused individuals verbalize intent early and maintain buffers; promotion-focused temperaments press harder for efficiency – effective on good days, brittle on bad ones (Wallace and Chen, 2006). Parallel findings outside aviation support this pattern: in a study of railway controllers, personality predicted whether operators sought buffers or took risks as situations flexed (Elfering et al., 2012). Across safety-critical domains, temperament influences how people perceive time pressure, uncertainty, and control.

Experience and Coordination in Action

With time, operators build **schemas** – mental shortcuts that compress procedures and free bandwidth. Experienced pilots don't rebuild a STAR from scratch; they draw on stored "if-then" patterns. Seasoned cabin managers know instantly which passengers will need help first in a diversion.

Yet, experience interacts with the traits above. A high-capacity, conscientious individual may detect drift early; a lower-capacity or impulsive operator may achieve speed but lose forward awareness. Hence, two people with identical flight hours can differ sharply once the plan changes late (Endsley, 2018).

A revealing example comes from **Delta Air Lines Flight 1086** in 2015. During a snowy landing at New York LaGuardia, the MD-88 began veering left after touchdown on the contaminated surface. As directional control degraded, both pilots made control inputs: the captain on the controls, while the first officer was operating thrust reversers and monitoring braking. For several seconds these actions were not fully coordinated, briefly compounding the drift. Once the captain called, "I have it," coordination tightened and the aircraft came to rest with its nose against the perimeter fence. The NTSB (2016) cited runway contamination and coordination gaps during rollout. The human factor was subtle but clear: under stress, even complementary skills can conflict unless role clarity is maintained. When recognized early and bridged through clear calls and mutual awareness, those same differences become a strength.

Implications for Selection and Training

Aviation has long shown that cognitive and psychomotor tests predict early training outcomes (Carretta, 2013; Carretta and Ree, 2003). The next step is to carry these differences into training design instead of assuming they fade in the simulator. CBTA and EBT expect instructors to observe how trainees manage bandwidth and SA, not just whether they complete the maneuver (EASA, 2020; ICAO, 2020;). The KSA100 framework extends this to ground school, requiring reasoning, projection, workload management, and metacognition to be assessed rather than assumed (EASA, 2018).

The goal is not to label people but to equip them with tools that fit their cognitive fingerprint.

Treat individual differences as design inputs. Choose coping methods that fit the person, the cockpit, the cabin, or the control room. Then SA stops relying on good days – it becomes something that can be built deliberately and sustained reliably.

MEASURING WORKLOAD AND SITUATIONAL AWARENESS

There are different ways we measure these concepts in aviation. Some rely on what operators report about their own experience; others observe behavior, physiology, or performance patterns. None are perfect, but together they have transformed how we understand human workload and awareness in complex, high-risk environments.

Why Measurement Matters

Workload and SA cannot be seen directly, yet both define how safely and effectively operations unfold. For decades, human factors specialists have tried to capture these invisible states, often imperfectly, but always with one goal: to make training evidence-based rather than impression-based. Without measurement, feedback relies on subjective comments like "looked busy" or "seemed fine," offering little insight into what was really happening inside the operator's head.

Understanding how to measure workload and SA matters because it shapes how we teach, coach, and design systems. Good tools reveal when the mental picture starts to thin or when attention collapses under stress. They give instructors and crews a shared language for what is otherwise invisible – the moment when performance begins to drift.

Subjective and Objective Approaches

Workload and SA cannot be seen directly, but both leave traces in what people feel, say, and do. Measurement methods fall broadly into two categories: subjective and objective. Each reveals part of the picture, and together they form a useful mirror of performance.

Subjective measures capture the operator's internal view. Rating tools or quick verbal probes ask how demanding a task feels in real time or shortly afterward (Hart, 2006). They are simple, flexible, and provide insight into the operator's mental state. Yet accuracy varies. Under pressure or social expectation, people often misjudge their own limits (Matthews et al., 2015). Pilots may downplay strain, controllers may normalize high tempo, and cabin crew might report feeling "fine" after challenging sectors.

Self-assessment works best when paired with external data. Over time, this pairing helps operators develop metacognitive awareness – the skill of sensing personal thresholds before performance slips. Many begin noticing early cues like tense breathing, muscle stiffness, or slower speech. Recognizing these signs becomes a self-regulation tool, not just a reflection.

Objective measures rely on observable behavior and physiology. As cognitive demand increases, eye movements narrow, fixations lengthen, and speech slows

(Wilson and Russell, 2003). Physiological indicators such as heart-rate variability, pupil dilation, and electrodermal activity also shift with rising effort (Hancock and Desmond, 2001). In air traffic control, higher complexity produces less systematic scanning and more corrective glances (Loft et al., 2007); in maintenance, elongated pauses or error-check loops often reveal mental overload.

The challenge is practicality. Tools must be accurate without intruding on the task itself. A headset or sensor that adds distraction defeats the purpose. The most useful systems are low-profile and reliable in real conditions, capturing enough data to guide coaching or debriefs without turning the operator into a test subject.

Together, subjective and objective measures help transform invisible mental effort into visible learning. They remind us that awareness is not guessed—it is observed, reflected, and trained.

From Data to Debrief

The value of any measurement lies in what it teaches. Numbers alone do not improve performance; insight does. Effective tools translate results into observations that instructors can act on: "Here is where your scan collapsed," or "This is where your projection stopped matching the situation." These moments turn abstract data into concrete coaching.

In competency-based frameworks the goal is not to produce perfect workload scores, but to sharpen awareness and judgment. Instructors gain the most when they can align an operator's perception of effort with observed behavior – for example, asking "How loaded were you here?" right after the task. When cross-checked, these perspectives show whether awareness and performance stayed synchronized.

At the organizational level, aggregated data reveal patterns that individuals cannot see. If many crews show rising workload at the same procedural point – for example, a late runway change or a handover between teams – that becomes a flag for redesigning procedures or reinforcing that scenario in training.

The best measurement tools close the feedback loop quickly. They let instructors and crews turn complex data into operational lessons that improve the very next flight or shift.

The Future of Workload and Situational Awareness

The next frontier in understanding workload and SA is prediction rather than description. After years of relying on observation and debriefs, aviation is moving toward real-time insight – helping people recognize strain before performance breaks down. The goal is not control, but partnership: tools that support human awareness rather than replace it.

Adaptive Systems – Sensing Strain before It Shows

Flight decks and control centers already produce rich data on control inputs, scan traces, and speech timing. Machine-learning tools can now translate these patterns into early signs of overload or fading SA. Adaptive automation, once a NASA experiment, is becoming reality: systems that quietly emphasize what matters when attention narrows (Byrne and Parasuraman, 1996; Kohn et al., 2021). But design must remain transparent. As Hancock and Desmond (2001) noted, support that feels like surveillance adds noise instead of help. A good system should act as a calm partner, not a supervisor.

Immersive and Bio-Adaptive Training

Virtual and augmented reality (VR/AR) make it possible to practice "safe overload" – an approach full of distractions, or "safe underload," such as quiet night shifts where vigilance drifts. Properly designed scenarios let crews rehearse recognition and recovery of thinning SA (İnan et al., 2024). Biofeedback extends this learning further. Subtle cues from heart-rate variability or pupil dilation teach operators to notice their own thresholds before errors appear (Prinsloo et al., 2014) The idea is simple: learn to read yourself as well as your instruments.

Design and Organization – The Quiet Safeguards

The future is not just technical. Clearer interfaces, balanced staffing, and thoughtful rostering often preserve awareness better than any sensor or algorithm (Endsley and Jones, 2012).

The future of conscious skies will rely less on perfect automation and more on humans and systems learning together – each aware of the other, adaptive, and ready to protect the fragile edge of awareness where safety lives.

TAKEAWAYS

For Pilots and Cabin Crew

- Anchor priorities when workload spikes: aviate-navigate-communicate, and make clear what will wait.
- Use light forward projections: short statements ("expect late downwind," "cabin secure in five") keep everyone in sync.
- Re-anchor vigilance in quiet phases with micro-checks or what-if scenarios to prevent drift.

Reflective Question: When was the last time you noticed your SA thinning – and what early cue could you watch for next time?

For Instructors, Examiners, and Leaders

- Observe how trainees cope under both overload and underload; debrief not only mistakes but how they managed bandwidth.
- Use scenario variety: pair high-load events with quiet stretches to expose both ends of the Yerkes-Dodson curve.

- Coach strategy fit: help sequential vs. multi-tasking trainees refine approaches rather than forcing one "best" style.

Reflective Question: *Do your feedback sessions capture "how they managed the picture" – or just whether they flew the profile?*

For Operational Teams

- Embed SA and workload drills into syllabi – not as abstract lectures but as designed moments in flight, cabin, and ATC training.
- Use measurement tools pragmatically: pair subjective self-ratings with behavioral or physiological cues to sharpen coaching.
- Audit procedures and rostering: look for recurring workload peaks or vigilance gaps that point to redesign, not just retraining.

Reflective Question: *At the organizational level, do we treat SA as an individual skill, or as a system property shaped by training, design, and staffing?*

REFERENCES

Bailey, B. P., Konstan, J. A. and Carlis, J. V. (2001) 'The effects of interruptions on task performance, annoyance, and anxiety in the user interface', *Proceedings of INTERACT 2001: Eighth IFIP TC.13 Conference on Human-Computer Interaction*, Tokyo, Japan, July 9–13, 2001. Amsterdam: IOS Press, pp. 593–601.

Bainbridge, L. (1983) 'Ironies of automation', *Automatica*, 19(6), pp. 775–779. https://doi.org/10.1016/0005-1098(83)90046-8

Beus, J. M., Dhanani, L. Y. and McCord, M. A. (2015) 'A meta-analysis of personality and workplace safety: Addressing unanswered questions', *Journal of Applied Psychology*, 100(2), pp. 481–498. https://doi.org/10.1037/a0037916

Byrne, E. A. and Parasuraman, R. (1996) 'Psychophysiology and adaptive automation', *Biological Psychology*, 42(3), pp. 249–268. https://doi.org/10.1016/0301-0511(95)05161-9

Carretta, T. R. (2013 'Predictive validity of pilot selection instruments for remotely piloted aircraft training outcome', *Aviation, Space, and Environmental Medicine*, 84, 47–53. https://doi.org/10.3357/asem.3441.2013

Carretta, T. R. and Ree, M. J. (2003) 'Pilot selection methods', in P. S. Tsang, M. A. Vidulich (Eds.), *Principles and Practice of Aviation Psychology*. Mahwah, NJ: Lawrence Erlbaum, pp. 357–396.

Christian, M. S., Bradley, J. C., Wallace, J. C. and Burke, M. J. (2009) 'Workplace safety: A meta-analysis of the roles of person and situation factors', *Journal of Applied Psychology*, 94(5), pp. 1103–1127. https://doi.org/10.1037/a0016172

Clarke, S. and Robertson, I. T. (2005) 'A meta-analytic review of the Big Five personality factors and accident involvement in occupational and non-occupational settings', *Journal of Occupational and Organizational Psychology*, 78(3), pp. 355–376. https://doi.org/10.1348/096317905X26183

Cooke, N. J., Gorman, J. C. and Winner, J. L. (2007) 'Team cognition', in E. Salas, G. F. Goodwin, C. S. Burke (Eds.), *Team Effectiveness in Complex Organizations: Cross-Disciplinary Perspectives and Approaches*. New York: Routledge

Diamond, D. M., Campbell, A. M., Park, C. R., Halonen, J., & Zoladz, P. R. (2007). 'The temporal dynamics model of emotional memory processing: A synthesis on the neurobiological basis of stress-induced amnesia, flashbulb and traumatic memories, and the Yerkes–Dodson law'. *Neural Plasticity*, Article ID 60803, 33 pages. https://doi.org/10.1155/2007/60803

Dismukes, R. K., Berman, B. A. and Loukopoulos, L. D. (2007) *The Limits of Expertise: Rethinking Pilot Error and the Causes of Airline Accidents*. Abingdon: Routledge. https://doi.org/10.4324/9781315238654

Durso, F. T. and Dattel, A. R. (2006) 'SPAM: The real-time assessment of situation awareness', in S. Banbury, S. Tremblay (Eds.), *A Cognitive Approach to Situation Awareness: Theory and Application*. Aldershot: Ashgate, pp. 137–154.

Elfering, A., Grebner, S. and Haller, M. (2012) 'Railway-controller-perceived mental work load, cognitive failure and risky commuting', *Ergonomics*, 55(12), pp. 1463–1475. https://doi.org/10.1080/00140139.2012.718802

Endsley, M. R. (1995) 'Toward a theory of situation awareness in dynamic systems', *Human Factors*, 37(1), pp. 32–64. https://doi.org/10.1518/001872095779049543

Endsley, M. R. (2018) 'Expertise and situation awareness', in K. A. Ericsson, R. R. Hoffman, A. Kozbelt, A. M. Williams (Eds.), *The Cambridge Handbook of Expertise and Expert Performance* (2nd ed., pp. 714–741). Cambridge: Cambridge University Press. https://doi.org/10.1017/9781316480748.037

Endsley, M. R. and Bolstad, C. A. (1994) 'Individual differences in pilot situation awareness', *International Journal of Aviation Psychology*, 4(3), pp. 241–264. https://doi.org/10.1207/s15327108ijap0403_3

Endsley, M. R. and Jones, W. M. (2012) *Designing for Situation Awareness: An Approach to User-Centered Design*. 2nd ed. Boca Raton, FL: CRC Press.

Engle, R. W. (2018) *Working Memory and Executive Attention: A Cognitive Neuroscience Perspective*. New York: Routledge.

European Union Aviation Safety Agency (EASA) (2018) *ED Decision 2018/001/R – Acceptable Means of Compliance and Guidance Material to Part-FCL (KSA100 Framework)*. Cologne: EASA.

European Union Aviation Safety Agency (EASA). (2020) *Easy Access Rules for Flight Crew Licensing (Part-FCL)*. Cologne: European Union Aviation Safety Agency. Available at: https://www.easa.europa.eu/sites/default/files/dfu/Easy_Access_Rules_for_Part-FCL-Aug20.pdf

European Union Aviation Safety Agency (EASA) (2023) *Evidence-Based and Competency-Based Training and Assessment (EBT/CBTA) Implementation Manual*. Cologne: EASA. Available at: https://www.easa.europa.eu/en/domains/aircrew-and-medical/evidence-based-training-ebt

Federal Aviation Administration (FAA) (2016) *Human Factors in Aviation Maintenance and Inspection (HFAMI) – Chapter 5: Workload and Fatigue*. Washington, DC: US Department of Transportation, Federal Aviation Administration. Available at: https://www.faa.gov/about/initiatives/maintenance_hf

Hancock, P. A. and Desmond, P. A. (Eds.) (2001) *Stress, Workload, and Fatigue*. Mahwah, NJ: Lawrence Erlbaum Associates Publishers.

Hart, S. G. (2006) 'Nasa-task load index (NASA-TLX); 20 years later', *Proceedings of the Human Factors and Ergonomics Society Annual Meeting*, 50(9), pp. 904–908. https://doi.org/10.1177/154193120605000909

İnan, M., Aydin, O., Yildirim, S. and Çelik, M. (2024) 'Enhancing situational awareness in aviation through eye-tracking and physiological data: A human factors approach', *Aerospace Science and Technology*, 152, p. 109593. https://doi.org/10.1016/j.ast.2024.109593

International Civil Aviation Organization (ICAO) (2020) *Procedures for Air Navigation Services – Training (PANS-TRG), Doc 9868, 3rd ed., Incorporating Amendments Approved by Council to 8 June 2020*. Montreal: International Civil Aviation Organization.

Available at: https://www.icao.int/sites/default/files/APAC/RASG/SafetyTools/19-Procedures-for-Air-Navigation-Services-Training-PANS-TRG-Doc-9868.pdf

Jafari, M.-J., Zaeri, F., Jafari, A. H., Payandeh Najafabadi, A. T. and Hassanzadeh-Rangi, N. (2019) 'Human-based dynamics of mental workload in complicated systems', *EXCLI Journal*, 18, pp. 501–512. https://doi.org/10.17179/excli2019-1372

Kane, M. J. and Engle, R. W. (2002) 'The role of prefrontal cortex in working-memory capacity…', *Psychonomic Bulletin & Review*, 9(4), pp. 637–671. https://doi.org/10.3758/BF03196323

Kohn, S., de Visser, E., Wiese, E. and Lee, Y.-C. (2021) 'Measurement of trust in automation: A narrative review and reference guide', *Frontiers in Psychology*, 12, p. 604977. https://doi.org/10.3389/fpsyg.2021.604977

Lievens, F., Peeters, H. and Schollaert, E. (2008) 'Situational judgment tests: A review of recent research', *Personnel Review*, 37(4), pp. 426–441. https://doi.org/10.1108/00483480810877598

Loft, S., Sanderson, P. M., Neal, A. and Mooij, M. (2007) 'Modeling and predicting mental workload in en route air traffic control: Critical review and broader implications', *Human Factors*, 49(3), pp. 376–399. https://doi.org/10.1518/001872007X197017

Mackworth, N. H. (1948) 'The breakdown of vigilance during prolonged visual search', *Quarterly Journal of Experimental Psychology*, 1(1), pp. 6–21. https://doi.org/10.1080/17470214808416738

Masi, D., Pini, A., Pozzi, S. and Rinaldi, S. (2023) 'Mental workload, stress and performance: A systematic review of workload assessment in safety-critical domains', *International Journal of Industrial Ergonomics*, 93, p. 103456. https://doi.org/10.1016/j.ergon.2023.103456

Mathieu, J. E., et al. (2000) 'The influence of shared mental models on team process and performance', *Journal of Applied Psychology*, 85(2), pp. 273–283. https://doi.org/10.1037/0021-9010.85.2.273

Matthews, G., Reinerman-Jones, L. E., Barber, D. J. and Abich, J. (2015) 'The psychometrics of mental workload: Multiple measures are sensitive but divergent', *Human Factors*, 57(1), pp. 125–143. https://doi.org/10.1177/0018720814539505

Morgan, B., D'Mello, S., Abbott, R., Radvansky, G., Haass, M. and Tamplin, A. (2013) 'Individual differences in multitasking ability and adaptability', *Human Factors*, 55(4), pp. 776–788. https://doi.org/10.1177/0018720812470842

Mulay, R. V., Gangwal, A., Shyam, A. K. and Sancheti, P. (2019) 'Prevalence and risk factors for work-related musculoskeletal disorders in flight attendants', *International Journal of Community Medicine and Public Health*, 6(6), pp. 2456–2460. https://doi.org/10.18203/2394-6040.ijcmph20192304

National Transportation Safety Board (2016) *Runway excursion during landing, Delta Air Lines flight 1086, McDonnell Douglas MD-88, N909DL, New York, March 5, 2015.* Aircraft Accident Report AAR-16/02. Washington, DC: NTSB. https://www.ntsb.gov/investigations/accidentreports/reports/aar1602.pdf

National Transportation Safety Board (2018) *Incident: Air Canada flight 759, Airbus A320-211, C-FKCK, San Francisco, California, July 7, 2017.* Aircraft Incident Report AAR-18/01. Washington, DC: NTSB. Available at: https://www.ntsb.gov/investigations/accidentreports/reports/air1801.pdf

Parasuraman, R. and Hancock, P. A. (2001) 'Adaptive control of mental workload', in P. A. Hancock, P. A. Desmond (Eds.), *Stress, Workload, and Fatigue: Theory, Research, and Practice.* Mahwah, NJ: Lawrence Erlbaum Associates, pp. 305–320

Parasuraman, R., Molloy, R. and Singh, I. L. (1993) 'Performance consequences of automation-induced 'Complacency'', *International Journal of Aviation Psychology*, 3(1), pp. 1–23. https://doi.org/10.1207/s15327108ijap0301_1

Pereira da Silva, M. F. (2014) 'Mental workload, task demand and driving performance: What relation?', *Procedia – Social and Behavioral Sciences*, 162, pp. 310–319. https://doi.org/10.1016/j.sbspro.2014.12.212

Prinsloo, G. E., Rauch, H. G. L. and Derman, W. E. (2014) 'A brief review and clinical application of heart rate variability biofeedback in sports, exercise, and rehabilitation medicine', *Frontiers in Psychology*, 5, 731. https://doi.org/10.3810/psm.2014.05.2061

Sarter, N. B. and Woods, D. D. (1995) 'How in the world did we ever get into that mode? Mode error and awareness in supervisory control', *Human Factors*, 37(1), pp. 5–19. https://doi.org/10.1518/001872095779049516

Stanton, N. A., Salmon, P. M., Walker, G. H., Jenkins, D. P. and Rafferty, L. A. (2009) *Human Factors Methods: A Practical Guide for Engineering and Design*. 2nd ed. Farnham: Ashgate.

Wallace, J. C. and Chen, G. (2006) 'A multilevel integration of personality, climate, self-regulation, and performance', *Personnel Psychology*, 59(3), pp. 529–557. https://doi.org/10.1111/j.1744-6570.2006.00046.x

Warm, J. S., Parasuraman, R. and Matthews, G. (2008) 'Vigilance requires hard mental work and is stressful', *Human Factors*, 50(3), pp. 433–441. https://doi.org/10.1518/001872008X312152

Wickens, C. D. (2002) 'Multiple resources and performance prediction', *Theoretical Issues in Ergonomics Science*, 3(2), pp. 159–177. https://doi.org/10.1080/14639220210123806

Wickens, C. D. (2008) 'Multiple resources and mental workload', *Human Factors*, 50(3), pp. 449–455. https://doi.org/10.1518/001872008X288394

Williamson, A., Lombardi, D. A., Folkard, S., Stutts, J., Courtney, T. K. and Connor, J. L. (2011) 'The link between fatigue and safety', *Accident Analysis & Prevention*, 43(2), pp. 498–515. https://doi.org/10.1016/j.aap.2009.11.011

Wilson, G. F. and Russell, C. A. (2003) 'Real-time assessment of mental workload using psychophysiological measures and artificial neural networks', *Human Factors*, 45(4), pp. 635–644. https://doi.org/10.1518/hfes.45.4.635.27088

Yerkes, R. M. and Dodson, J. D. (1908) 'The relation of strength of stimulus to rapidity of habit formation', *Journal of Comparative Neurology & Psychology*, 18, pp. 459–482. https://doi.org/10.1002/cne.920180503

Young, M. S., Brookhuis, K. A., Wickens, C. D. and Hancock, P. A. (2015) 'State of science: mental workload in ergonomics', *Ergonomics*, 58(1), pp. 1–17. https://doi.org/10.1080/00140139.2014.956151

7 Decision-Making and Problem-Solving

Navigating Complexity

CASE STUDY: INDONESIA AIRASIA FLIGHT QZ8501 (2014) – THE DRIFT INTO TRAGEDY

It was meant to be an ordinary Sunday flight across Southeast Asia. On December 28, 2014, *Indonesia AirAsia Flight QZ8501*, an Airbus A320-216 (PK-AXC), pushed back from Surabaya bound for Singapore. One hundred and fifty-five passengers and seven crew filled the cabin. Over the Java Sea, convective weather was already building, towering cumulonimbus clouds typical of the Intertropical Convergence Zone, the kind of tropical turbulence that keeps even seasoned crews alert.

During the climb through 32,000 feet, a small amber message appeared on the Electronic Centralized Aircraft Monitor (ECAM): *Rudder Travel Limiter System Fault*. It was a familiar nuisance. The same fault had been recorded 23 times in the preceding year, frustrating engineers and crew alike. To the pilots that morning, it appeared similar to previous occurrences. But familiarity can be dangerous. Instead of following the Airbus abnormal procedure, the captain pulled the FAC (flight augmentation computer) circuit breaker – an action prohibited in flight.

Within seconds, the autopilot and autothrust disconnected, returning the aircraft to manual control amid turbulence and distraction. The first officer was flying. His initial inputs were abrupt and nose-high: control inputs were abrupt and nose-high. The aircraft pitched up steeply, trading airspeed for altitude until energy bled away. The stall warning sounded continuously, but in the escalating workload, the crew did not effectively recognize and respond to the situation as a stall.

The captain, engaged in resetting the circuit breakers, rejoined the control of the aircraft as the situation deteriorated. Communication fragmented: "Pull down, pull down" overlapped with "I'm trying." Neither verbalized the word *stall*. For more than three minutes, the A320 remained fully stalled, descending through cloud until it impacted the Java Sea with no survivors (KNKT, 2015).

The National Transportation Safety Committee of Indonesia (KNKT, 2015) concluded that this was not a single catastrophic error but a chain of small, routine misjudgments: frustration, procedural deviation, divided attention, loss of situational awareness, and incorrect decisions. Familiarity with the recurring fault influenced how the situation was perceived and managed.

DOI: 10.1201/9781003515418-7

INTRODUCTION

Decision-making and problem-solving in aviation is never abstract – it is survival in action. Checklists, Standard Operating Procedures (SOPs), and automation all exist to channel choices toward safety, yet as QZ8501 showed, even well-trained crews can drift into patterns that quietly steer them away from good outcomes.

The failure rarely comes from ignorance or lack of discipline but from how the human brain handles pressure. When time compresses and uncertainty grows, thought narrows to what feels familiar. Heuristics, so mental shortcuts that usually help, become traps when they disguise new hazards as old ones. A fault that "looks like last time" may in fact be something very different.

Information in the cockpit rarely arrives cleanly. It comes in fragments: a chime, a message line, a radar echo, a passenger report. The hardest calls are often made when emotions are high and time is short. Research summarized in *Decision Making in Aviation* (Harris and Li, 2015) shows that under pressure, pilots tend to rely on recognition-primed decision-making, matching patterns from experience rather than conducting full analytical reasoning. Martinussen and Hunter (2017) found that stress and workload narrow attention, while Flin et al. (2008) highlighted that sound judgment begins with recognizing that a decision is needed at all.

As focus tightens, that recognition can vanish. Problem-solving without reflection becomes problem-reacting, and in aviation, that difference can decide the outcome (Flin et al., 2008). The Indonesia AirAsia tragedy illustrated this vividly: both pilots were operating inside separate cognitive tunnels, each convinced he understood the situation (Harris and Li, 2015). Understanding how such choices unfold allows us to design training, systems, and cultures that support clear thinking, not just in rare emergencies but in the flow of everyday operations.

THE HUMAN BRAIN AND DECISION-MAKING

Decisions in aviation are made by the same brain that evolved for survival, not for complexity. Understanding how that brain operates, how it processes information, reacts to stress, and shortcuts analysis, is the first step to improving judgment across any operational role.

Dual-System Thinking

The human brain runs on two systems of thought (Kahneman, 2011).

- **System 1** is fast and intuitive, drawing on experience to make instant judgments. It is what allows a pilot to sense a drift, a cabin crew member to calm a passenger before words are spoken, or an engineer to detect a mechanical irregularity by sound alone.
- **System 2** is slower and deliberate. It weighs alternatives, challenges assumptions, and applies reasoning. It takes energy and concentration, which makes it fragile under stress.

In everyday operations, System 1 handles most tasks efficiently. But when events turn ambiguous, instinct alone can mislead. Under fatigue or time pressure, people tend to rely on past patterns even when the context has changed (Martinussen and Hunter, 2017). That is exactly what happened in QZ8501: a reflexive "fix" based on past experience replaced deliberate diagnosis.

Neurobiology of Decision-Making under Stress

Stress and fatigue do more than cloud judgment – they recalibrate the brain's decision machinery. Under acute stress, valuation and learning systems become biased toward immediate rewards and familiar patterns, accelerating reactions but undermining balanced analysis (Porcelli and Delgado, 2017).

Biochemically, stress floods the brain with **dopamine and cortisol**, heightening focus yet distorting how information is prioritized. Attention locks onto cues that seem most urgent, while subtler or conflicting data fade from view. At the same time, **glucose metabolism in the frontal lobes** decreases, reducing the energy available for working memory and integration of complex data. Fatigue emphasizes these changes: as metabolic efficiency drops, decision speed may remain high, but the *quality* of reasoning falls (Caldwell and Caldwell, 2016).

The paradox is striking. A pilot may feel acutely alert, even energized, while in reality perceiving less and processing more narrowly. Checklist items are skipped, warnings blend into background noise, and confirmation bias strengthens its grip. These are not lapses of skill or discipline, but predictable neurobiological outcomes of a brain temporarily tuned for survival rather than strategy. In aviation, recognizing that shift is part of mastering it.

THE DECISION ENVIRONMENT: PRESSURES, HIERARCHIES, AND TRAPS

Understanding why pilots make certain decisions means looking beyond the individual and into the **environment** that shapes them. In Indonesia AirAsia Flight 8501, fatigue, cockpit hierarchy, incomplete information, and the sudden loss of automation all combined to push two competent pilots into a corner where good judgment slipped away. These forces didn't act in isolation – they built on one another, layer by layer, until the margin for sound decision-making disappeared.

In this section we will break down the most powerful contextual forces that shape pilot judgment – fatigue and time pressure, authority gradients, ambiguity, and automation – showing how each played a role in QZ8501 and how similar dynamics have appeared across aviation.

Fatigue, Stress, and Time Pressure

Decision quality collapses when the clock feels louder than the mind. Fatigue drains mental flexibility, stress narrows the scan, and time pressure tempts people to close options too early. As discussed in *Chapter 5: Battle for Alertness: Attention, Fatigue, and Sleep*, tiredness and stress combine to shrink working memory until key cues vanish.

In Indonesia AirAsia 8501, the early departure and repeated rudder fault created just such an environment. The crew felt an urgency to fix the issue quickly, and that urgency narrowed their mental bandwidth. Research confirms this pattern – under combined fatigue and stress, deliberate analysis becomes harder to sustain, and instinct fills the gap (Caldwell and Caldwell, 2016; Porcelli and Delgado, 2017)

Another accident, of **Air India Express Flight 812** in 2010, at Mangalore showed how fatigue and stress can quietly dismantle judgment. After an overnight flight from Dubai, the captain – who had been resting during part of the flight and was likely experiencing reduced alertness during the descent – attempted a steep, unstable approach and persisted despite multiple go-around cues, overrunning the runway. The Boeing 737 overran the runway and broke apart, killing 158 people. Investigators concluded that cumulative fatigue, sleep inertia, reduced alertness and plan continuation bias under time pressure led to the decision to persist with the approach rather than go around (DGCA, 2010)

In such conditions, decisions are rarely reckless in intent, but constrained by a shrinking cognitive window, as fatigue, stress, and time pressure narrow attention and make abandoning an established plan increasingly difficult.

The Role of Authority Gradient

Cockpit decisions are never made in isolation – they are shaped by hierarchy, culture, and personality. Authority gradient describes the psychological distance between captain and first officer, and it can either encourage open dialogue or stifle it.

In F8501, communication broke down at the point of greatest need. The first officer's (FO's) excessive inputs went unchallenged, and when the captain returned to the cockpit, leadership was not firmly re-established. Helmreich (2000) warned that unchecked authority gradients can block the flow of critical information and leave crews working in parallel rather than as a unit.

This danger is not confined to the past. In 2022, **LATAM Airlines Flight 2213**, an *Airbus A320neo* (CC-BHB) departing Lima for Juliaca, collided with a fire truck crossing the runway during takeoff. The aircraft was accelerating through high speed when three airport rescue vehicles entered the active runway without confirmed clearance. Despite the crew's attempt to rotate, the aircraft struck one of the trucks, killing two firefighters and severely damaging the jet.

The final investigation by the Peruvian Directorate General of Civil Aviation (DGAC Peru, 2023) identified serious coordination and communication failures between air traffic control, airport operations, and rescue services. Ambiguity over operational authority and misaligned assumptions about clearance led to poor decisions and simultaneous, conflicting actions on the same runway (DGAC, 2023).

While not a cockpit authority gradient case in the classic sense, the event exposed the same systemic vulnerability: hesitation to clarify, incomplete communication loops, and fragmented situational awareness across hierarchies, which influence proper decision-making. As discussed in *Chapter 8: Leadership and Teamwork: Coordinating the Invisible*, flattening these gradients and reinforcing assertive coordination across all operational roles remain key defenses against silence and assumption.

Environmental Conditions, Individual Perception and Illusions

Decision-making in aviation does not occur in a vacuum. It is shaped by the environment in which it unfolds – by light and darkness, acceleration and noise, workload and time pressure. Human perception is continuously calibrated by these external cues, and when they are degraded or misleading, the body's sensory systems can generate experiences that feel urgent and authoritative, even when they conflict with instrument information. In such conditions, perception itself becomes a critical element of the decision environment.

The human vestibular system evolved to support orientation and movement in a gravitational environment on the ground. In flight, particularly during rapid linear acceleration and in conditions of limited external visual references, this system becomes vulnerable to misinterpretation. The otolith organs respond identically to changes in gravitational force and to forward acceleration, making it physiologically impossible to distinguish between a pitch change and acceleration based on bodily sensation alone. This limitation underlies somatogravic illusion, a well-documented phenomenon in aviation that becomes most pronounced during take-off and go-around, especially at night or in instrument meteorological conditions (Gillingham & Previc, 1993).

The **Atlas Air Flight 3591** accident in 2019 illustrates how ambiguity can quickly distort situational understanding. A Boeing 767 operating a scheduled cargo flight, crashed near Trinity Bay, Texas, during descent toward Houston. The aircraft was operating in instrument meteorological conditions, with cloud cover eliminating external visual references during a high-workload phase of flight.

During the descent, an erroneous activation of go-around mode led to an increase in thrust and pitch. Flight-data records showed rapid acceleration accompanied by a pitch-up command. In the absence of reliable visual cues, the first officer likely perceived the aircraft's motion in a way that was inconsistent with instrument indications, potentially influenced by vestibular sensations rather than instrument information. This "seat-of-the-pants" perception – the body's intuitive sense of motion based on vestibular and proprioceptive cues (Stott, 2013) – can be compelling but is unreliable in flight, as the otolith organs cannot distinguish between forward acceleration and a pitch-up attitude.

The first officer responded with sustained nose-down control inputs, consistent with a misinterpretation of the aircraft's energy state and an illusion of excessive pitch or impending stall. In reality, the aircraft remained aerodynamically stable. The rapid control inputs, combined with degraded SA, sudden stress and time pressure, resulted in a steep dive from which recovery was not achieved.

The investigation found no evidence of mechanical failure. Instead, it concluded that the first officer's inappropriate control inputs, influenced by misinterpretation of aircraft state and automation behavior, led to a rapid breakdown in situational awareness and ineffective monitoring. The event illustrates how, in conditions of limited visibility and ambiguous feedback, intuitive "seat-of-the-pants" sensations can override instrument-based understanding, shaping decisions within seconds and with catastrophic consequences NTSB (2020).

This case illustrates how ambiguous feedback, particularly when mediated through automation, can rapidly distort a crew's understanding of what the aircraft is actually doing. In moments of startle and time pressure, crews may act on an incomplete or incorrect mental model, highlighting the importance of deliberate verification and shared sense-making when the situation begins to diverge from expectations (Dekker, 2014; Endsley, 1995; Klein, 2008).

Ambiguity is therefore not a neutral condition; it actively invites the brain to default to what feels familiar. Without intentions spoken out loud and structured cross-checking, those defaults can feel as certain as fact (Dekker, 2014).

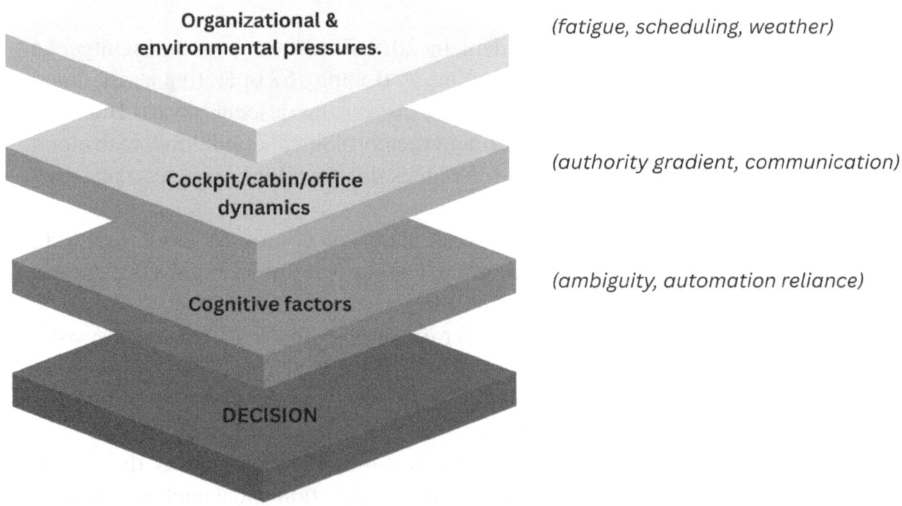

FIGURE 7.1 Layers of influence on aviation decisions.

Automation and Skill Decay

Automation shields crews from workload, but it also hides complexity. When it suddenly disconnects, the gap between expectation and reality can be brutal. In QZ8501, the autopilot's quiet disengagement left the crew facing manual flight in storm conditions while still preoccupied with the rudder fault. The loss of automatic stability turned a nuisance into a crisis.

This is the paradox of automation. It makes flight smoother and safer most of the time, yet weakens manual skills and situational resilience when it steps back. As we will explore in *Chapter 10: Manual Control, Automation, and Technology: Hands-on or Hands-off?* designing systems that support rather than erode human expertise remains one of aviation's toughest challenges.

These pressures rarely act in isolation – they build in layers. Figure 7.1 illustrates how organizational factors, team dynamics, and cognitive influences all combine to shape the final decision, whether in the cockpit, cabin, or office.

PROBLEM-SOLVING IN THE COCKPIT

Problem-solving in aviation is rarely about finding a solution. More often, it is about working out what the *real* problem is. Many accidents show that crews did not run out of options – they ran out of clarity. Structured models exist to help pilots frame the issue, but in the heat of the moment, they are not always used. Crews slide instead between intuitive "fast and frugal" reactions and slower, systematic reasoning.

Fixation and Over-analysis

In Flight 8501, the crew saw a familiar rudder fault and assumed it was manageable. They bypassed formal diagnosis and tried a workaround, applying an old mental script to a new situation. Psychologists call this **fixation** – a form of cognitive rigidity that keeps people locked into their first interpretation, even when new evidence points elsewhere (Wickens and McCarley, 2008).

The opposite problem can also arise. **Over-analysis** under pressure is its own hazard. In fast-moving events, waiting too long for perfect understanding can cost vital seconds. The challenge is balance: responding confidently, but not blindly (Orasanu and Martin, 1998). That balance becomes even harder in modern cockpits, where decision-making must be managed alongside automation modes, team communication, and standard procedures.

Structured Decision-Making Models

In high-stakes environments, structured decision-making models act as anchors. When stress, time pressure, and incomplete data push the brain toward instinctive reactions, these models slow the tempo just enough to re-engage deliberate, System 2 thinking. They turn fast, reactive moments into manageable steps.

Aviation has long relied on simple, easy-to-recall acronyms – compact mental checklists that keep reasoning on track when pressure rises. Among the most established are **FORDEC**, **T-DODAR**, and **DECIDE**, each designed to bring order and reflection into moments of uncertainty (CAA UK, 2002; FAA, 1991; Orasanu and Martin, 1998).

FORDEC – Facts, Options, Risks, Decision, Execution, Check

- *Facts:* What's happening? What do we actually know?
- *Options:* What possible courses of action exist?
- *Risks:* What are the trade-offs of each?
- *Decision:* Choose the option that best balances safety and outcome.
- *Execution:* Carry it out with discipline.
- *Check:* Monitor the result and adapt as needed.

T-DODAR – Time, Diagnose, Options, Decide, Act, Review

- *Time:* How much time do we really have?
- *Diagnose:* Define the issue clearly.
- *Options:* Generate possible responses.
- *Decide:* Select the best course.
- *Act:* Execute decisively.
- *Review:* Evaluate whether it worked.

DECIDE – Detect, Estimate, Choose, Identify, Do, Evaluate

- *Detect:* Notice that something has changed.
- *Estimate:* Judge its importance.
- *Choose:* Define the goal.
- *Identify:* Work out actions to achieve it.
- *Do:* Carry them out.
- *Evaluate:* Assess whether they succeeded.

Each model has its own emphasis: FORDEC balances analysis with execution, T-DODAR highlights time management, and DECIDE focuses on cognitive flow. But all share a vital feature: **a built-in feedback loop**. That loop is what keeps decision-making dynamic rather than static, inviting the crew to reassess assumptions, integrate new cues, and adjust actions before small deviations grow into larger risks. Good decisions are not one-time acts; they evolve as new information emerges (Flin et al., 2008).

From Knowing to Applying

Training aviation personnel in structured decision-making is not new. Since the 1990s, aviation psychologists have shown that **scenario-based learning** strengthens the ability to make effective choices under pressure (Orasanu and Martin, 1998; Salas et al., 2006, 2009). Crews exposed to realistic, high-stress simulations perform better at diagnosing problems and coordinating responses. Although knowledge alone is never enough, models only help when they become instinctive: shifting from acronyms to living habits in daily operations.

Decision-making tools must therefore live in the operational language of the cockpit, cabin, and operations center. **Guided coaching, instructor prompts,**

and structured debriefs reinforce their use until it feels natural, helping pilots and crew translate classroom theory into automatic behaviors (Flin et al., 2008; O'Connor et al., 2008). Instructors play a key role by encouraging their application during both normal and abnormal training so that, under stress, recall is effortless.

These models are not checklists but **scaffolds for judgment**. Crews flex between them depending on context: FORDEC when time allows, T-DODAR under urgency, DECIDE when the issue is unclear. What matters is not the acronym but engaging structured reasoning instead of bypassing it.

When rehearsed through realistic, role-specific scenarios: system failures, cabin medical events, ATC disruptions, or maintenance faults, these tools build the mental "muscle memory" that lets deliberate thought cut through instinct. In time, structured decision-making evolves from knowledge to culture – across cockpits, cabins, and control rooms alike.

TRAINING DECISION-MAKING AND PROBLEM-SOLVING

In aviation, clear and timely decisions prevent small issues from turning into crises. This applies equally to pilots managing system failures, cabin crew handling disruptive passengers, or dispatch staff dealing with weather changes. Good decision-making is not accidental – it must be deliberately developed.

Models like FORDEC, T-DODAR, and DECIDE offer useful scaffolds, but the real challenge is turning them into habits. Approved Training Organizations (ATOs), airlines, and training departments must help people use these tools in context, not as academic acronyms but as **practical methods** that support everyday judgment (CAA UK, 2002; FAA, 1991; Orasanu and Martin, 1998). The aim is not memorization but the integration of structured thinking into normal workflow, so that under stress or ambiguity, sound habits emerge automatically.

When reinforced through **realistic simulations, team scenarios, and timely feedback**, structured decision-making grows beyond procedure; it becomes part of operational culture (Salas et al., 2012).

EMBEDDING JUDGMENT THROUGH TRAINING

Structured decision-making cannot be taught through lectures alone. It requires immersive, experience-based learning that mirrors operational reality across roles.

- **Scenario-Based Learning:** Whether in a flight simulator, a cabin mock-up, or an operations control exercise, realistic scenarios encourage participants to apply decision models in real time. By adding ambiguous cues, compressed timelines, and evolving risks, training helps people practice diagnosing problems, weighing options, and acting decisively under pressure (Orasanu and Martin, 1998).
- **High-Fidelity Simulation:** Pilots benefit from full-flight simulators, but cabin crew, ATC staff, and maintenance teams also need high-fidelity environments. Fire drills, medical emergencies, or communication breakdowns

provide safe opportunities to test judgment. When instructors layer in distractions and failures, participants learn to balance instinct with structured reasoning (Hays et al., 1992).
- **CRM Integration:** Decision-making is never a solo act. Crew Resource Management (CRM) applies in the cockpit, the cabin, and the operations office. Shared mental models, assertiveness, and cross-checking ensure that decisions are not just fast but coordinated. Good training builds collaboration into the process, teaching people not only what to decide but how to decide as a team (Salas et al., 2006).

The most effective programs build decision-making not only for rare emergencies but as a continuous background process. Through repetition, guided reflection, and cross-role practice, structured judgment becomes second nature across the aviation system.

From Novice to Expert: Models of Progression

Becoming an effective decision-maker is a gradual process, whether as a cadet pilot, junior cabin crew member, or new operations controller. People don't leap from novice to expert – they move through stages where structure dominates early learning and intuition develops only after habits take root.

1. The **Dreyfus Model of Skill Acquisition** (Dreyfus and Dreyfus, 1980) describes five stages: *novice, advanced beginner, competent, proficient, and expert.* Novices rely on strict rules because they lack the context to adapt. With experience, judgment deepens, and by the expert stage, actions are guided by intuition shaped through thousands of exposures. This explains why a new first officer or dispatcher clings to rules, while experienced colleagues operate with greater fluidity.
2. The **Recognition-Primed Decision (RPD) Model** (Klein, 1998) shows how experts often bypass deliberate comparisons, instead recognizing patterns and mentally simulating outcomes before acting. A captain anticipating a diversion or a senior flight attendant sensing a medical escalation are using years of stored cognitive rehearsal that make decisions appear instant.
3. **Adaptive Expertise** (Hatano and Inagaki, 1986) adds a final layer: distinguishing routine expertise, which ensures efficiency, from adaptive expertise, which allows flexibility and innovation when standard scripts fail. Training that includes ambiguity, time pressure, and creative problem-solving builds this adaptive edge.

Together, these frameworks show that expertise is less about efficiency than adaptability. For instructors and line captains, recognizing where learners sit on this continuum helps tailor feedback structure and rule adherence for beginners, and exposure to uncertainty for those ready to develop intuitive and adaptive judgment.

The Path and Perspectives of Decision-Making Development

Structured and intuitive decision-making are not opposites – they are stages of the same developmental path (Figure 7.2). Early in training, **structure** dominates. Acronyms, checklists, and frameworks anchor judgment when experience is limited. With time, these habits move into the background, allowing intuition to emerge as a dependable tool (Kahneman, 2011; Klein, 1998).

From the **learner's perspective**, the shift feels like moving from dependence to independence. Decision-making begins slow and deliberate, guided by written rules or instructor prompts. As exposure grows, the same steps blend into instinctive habits. Confidence builds, patterns become easier to spot, and decisions feel natural rather than forced. Even then, structure remains as a safety net beneath intuition.

For instructors, pacing is critical. Early on, the goal is discipline, anchoring trainees in structured methods and preventing shortcuts that hide weak understanding. As learners advance, instructors introduce ambiguity, time pressure, and competing demands. These challenges create the space for intuition to develop. Debriefs then tie intuitive actions back to structured principles, showing that intuition is trustworthy only when built on solid training and reflection.

This dual pathway applies across aviation roles. A cadet mastering FORDEC, a new flight attendant handling a cabin emergency, or a controller managing a weather diversion all start with procedure and progress toward adaptability – flexible under pressure, but disciplined enough to return to structure when it matters most.

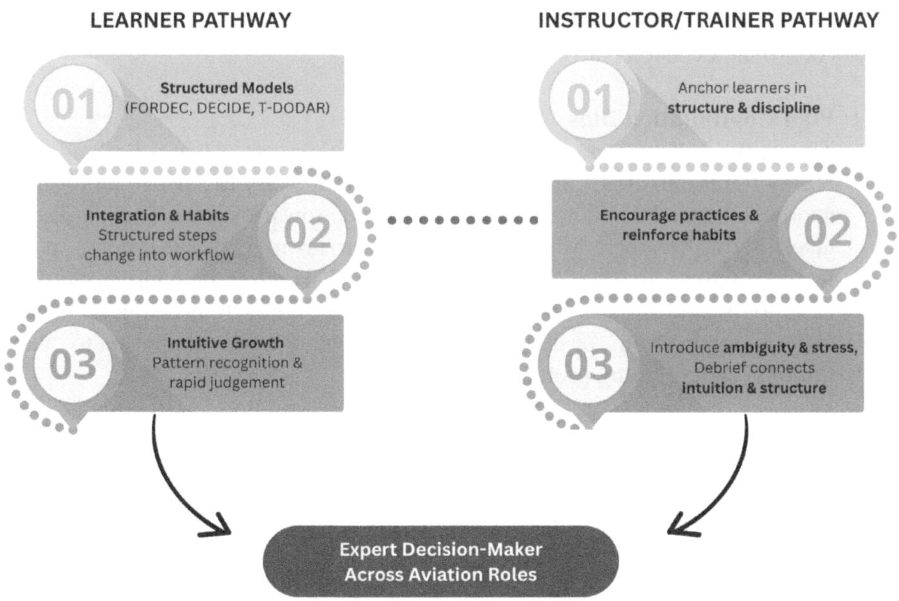

FIGURE 7.2 The path of decision-making development.

DECISION TRAPS, BIASES, AND THE PROBLEM-SOLVING BRAIN

Decision-making and problem-solving sit at the core of aviation. Crews don't just follow checklists – they interpret, adapt, and improvise when reality drifts from procedure. Yet the human brain, for all its power, is fallible.

Neuroscience shows that in high-stakes settings, the brain often reverts to fast, automatic processing to conserve effort (Kahneman, 2011). Under acute pressure, this shift becomes even stronger as stress hormones narrow focus and suppress reasoning (Leach, 2004). These mental shortcuts, known as **heuristics**, usually serve us well: they let a pilot manage a busy descent, a flight attendant respond to a medical emergency, or a controller reroute traffic efficiently (Wickens and McCarley, 2008). But when fatigue enters the picture, neural efficiency drops, and those same shortcuts can turn into **biases:** predictable thinking errors that distort what we see and how we act (Caldwell and Caldwell, 2016).

Biases affect not only decisions but also solutions. Crews may spot the right problem and yet pursue the wrong fix through assumption or overconfidence. Understanding these patterns (and recognizing their traces in real operations) helps pilots, instructors, and teams detect traps early and build a culture resilient to them.

Bias in Action: Real Aviation Examples

To explain the most common biases that creep into aviation decisions, it helps to associate them with some real events. Below are some of the most recognizable biases, linked to incidents where they played a role (some of the cases are explained in other parts of the book as well). While several biases may be present in a single event, the cases below are grouped according to the bias that strongly shaped the crew's dominant decision pathway.

1. **Plan Continuation Bias**
   This is the pull to press on with the original plan even when it no longer fits reality (Orasanu, 2002). It is why crews sometimes try to salvage an unstable approach instead of going around, or continue toward the destination despite deteriorating weather. Abandoning a plan carries psychological cost – it can feel like failure, inefficiency, or unnecessary disruption. In aviation, this bias commonly manifests as continuing an unstable approach, pressing on despite abnormal indications, or delaying a return decision while attempting to "work through" a developing problem.

Lion Air Flight 610 (2018): Shortly after departure, the crew experienced repeated uncommanded nose-down stabilizer trim inputs generated by the Maneuvering Characteristics Augmentation System (MCAS). Despite contradictory indications and an increasing workload, the crew did not declare an emergency or interrupt the flight by returning to the departure airport. Successive actions were taken within the assumption that the flight could still be continued safely, rather than reframing the situation as one requiring a change of plan. The accident investigation identified factors consistent with plan continuation bias as contributing to the crew's response, illustrating how commitment to an initial plan can delay decisive action even as evidence mounts that the situation has fundamentally changed(KNKT, 2019).

2. **Confirmation Bias**
   Here, we notice and value information that fits what we already believe, while sidelining evidence that contradicts it (Reason, 1990). Once an interpretation is formed, incoming cues are filtered to maintain coherence, particularly under time pressure or cognitive load.

EasyJet Flight 2146 (2019): During take-off from Lisbon, the crew of an Airbus A320 received a pitch-trim fault warning. Based on prior experience with similar warnings that had resolved without consequence, they interpreted the indication as benign and continued the take-off. Once airborne, increasing pitch-control stiffness confirmed that the fault was genuine, necessitating an emergency return. The investigation identified factors consistent with confirmation bias and normalization of deviance as contributing contributing to the crew's response, demonstrating how prior experience can shape interpretation in ways that delay recognition of a developing hazard (GPIAAF, 2020).

3. **Groupthink**
   Groupthink occurs when team members suppress doubts or alternative views in order to maintain harmony or follow authority (Janis, 1982). In aviation, this can prevent challenge, erode monitoring, and mask deteriorating performance until it is too late.

Aeroflot Flight 1492 (2019): Following a lightning strike shortly after take-off, the aircraft returned to Moscow and conducted a high-energy approach that culminated in a hard landing and post-impact fire. Although the First Officer was present and aware of abnormal parameters, effective challenge and intervention were limited, and monitoring responsibilities were not robustly shared. The investigation of Interstate Aviation Committee (IAC, also known as MAK) highlighted fragmented crew coordination, an authority gradient, and minimal cross-monitoring during a highly demanding phase of flight. The absence of assertive questioning and redistribution of control authority narrowed the crew's collective decision space (IAC, 2020), illustrating how groupthink can manifest not through overt agreement, but through what is left unspoken at critical moments.

4. **Overconfidence Bias**
   This is the trap of overestimating our ability to cope, which leads to risky actions or skipped steps (Kahneman, 2011). In aviation, this can manifest as pressing on despite degraded margins, accepting unstable conditions, or underestimating how rapidly a situation can escalate.

Pakistan International Airlines Flight 8303 (2020): During the approach into Karachi, the aircraft remained unstable, with excessive speed and altitude deviations. Despite multiple cues indicating an unstabilized approach and opportunities to discontinue the approach, the crew continued. The aircraft touched down with the landing gear retracted, resulting in engine damage; nevertheless, the crew initiated a go-around and attempted to continue the flight before ultimately crashing on a subsequent approach. The investigation identified poor judgement, continuation of an unsafe strategy, and inadequate threat assessment as contributing factors (AAIB Pakistan, 2020). This case illustrates how overconfidence in one's ability to "salvage" a situation can override objective indications that safety margins have already been exceeded.

5. **Hindsight Bias**
   After an accident, outcomes often appear obvious – "how could they not have seen it?" But in real time, the signals were rarely so clear. Hindsight bias makes learning harder because it distorts how difficult recognition actually was (Fischhoff, 1975).

Germanwings Flight 9525 (2015): In the aftermath of the crash, widespread commentary suggested that the pilot's mental health difficulties should have been apparent and preventable. However, the investigation showed that the available indicators were dispersed across time, organizations, and individuals, none of whom possessed a complete picture. Medical confidentiality, incomplete reporting, and the absence of overt behavioral cues meant that no single actor could clearly identify the impending risk (BEA, 2016). Viewed through hindsight, these fragments appear connected; viewed prospectively, they were isolated, ambiguous, and insufficient to trigger decisive intervention. The case illustrates how hindsight bias can oversimplify complex causal chains and lead to unrealistic expectations about detection and prevention.

Biases are not flaws of character – they are features of the human brain under pressure (Reason, 1990). But once we name them, we can design countermeasures: structured models, challenge-and-response culture, and debriefs that surface hidden assumptions. Aviation history shows what happens when biases run unchecked. Our task is not only to remember these lessons, but to embed them into everyday resilience.

MONITORING AND MITIGATION IN REAL TIME

Even the most skilled aviation professionals can slip into faulty reasoning under pressure. That is why decisions and actions must be monitored in real time, not only by the individual but by the team around them. Safety is not about avoiding mistakes; it is about detecting, challenging, and recovering from them early, before they cascade into something irreversible.

This principle applies across aviation. On the flight deck, in the cabin, at ATC consoles, and in engineering bays, the ability to notice, question, and intervene defines a true safety culture. Whether issuing a clearance, briefing a drill, or inspecting a worn component, shared vigilance turns isolated expertise into collective protection.

Building a Monitoring Culture

In high-performing teams, monitoring is not micromanagement—it is collaboration. Mutual awareness and shared responsibility keep systems resilient (Flin et al., 2008). Cabin crew may notice subtle passenger cues, Air Traffic Control Officers (ATCOs) catch deviations early, and engineers spot small inconsistencies before they become faults. Everyone contributes by noticing, questioning, and speaking up.

The harder part is voicing concern. Rank, self-doubt, or fear of conflict often silence the voice of caution. Yet silence remains risk's closest ally.

One painful lesson came with **Turkish Airlines Flight 1951** in 2009. On approach to Amsterdam, a faulty radio altimeter caused the autothrottle to reduce thrust to idle, leading to a gradual decay in airspeed. The monitoring pilot became aware of the developing problem, but his callouts remained limited and did not translate into assertive intervention. Monitoring, in this case, stopped at observation. The captain, operating under the assumption that the automation was managing speed appropriately, did not respond in time. The aircraft stalled short of the runway, killing nine people. The Dutch Safety Board concluded that while cues were available, the monitoring role had not been converted into effective action or challenge, allowing the situation to deteriorate unchecked (Dutch Safety Board, 2010).

Monitoring culture also means proactive detection, not just post-event callouts. **Flight Data Monitoring (FDM)** and **Flight Operational Quality Assurance (FOQA)** programs can identify unsafe trends before they cause incidents. ICAO (2018) and the FAA (2010) highlight how structured feedback loops from FDM data reduce event frequency – especially when findings are shared within a just culture that empowers crews to act.

Structured debriefs, cabin-to-cockpit reporting, and CRM training strengthen mutual vigilance (Salas et al., 2006). ATCOs and cabin crew often spot early warning signs before pilots do – if the system encourages them to speak.

Recognizing Cognitive Overload

As discussed in *Chapter 5: Battle for Alertness: Attention, Fatigue, and Sleep*, cognitive overload reduces awareness and decision quality. Here, the question is operational: how can we detect it in real time and recover before safety is compromised?

The signs are visible if we know where to look. A colleague who becomes quiet, stares at one display, or misses cues may already be at the limit. Overload feels like tunnel vision, fixation, or loss of the bigger picture. It affects every role: cabin crew can miss early medical signs, engineers slip into "tick-box" mode, and ATCOs delay conflict resolution as demands pile up.

Avoiding overload entirely is unrealistic, but early intervention works. Simple tools help:

- **Safety phrases** ("Let's double-check that," "Are we missing anything?") surface concern without confrontation.
- **Task redistribution** – one pilot taking radios, a crew member stepping aside, or an ATCO zooming out, creates cognitive space.
- **Grounding techniques**, like a deliberate breath or widening one's scan, reset attention and re-engage deliberate (System 2) thinking (Kahneman, 2011).

Recognizing overload is a shared responsibility. Individuals must self-monitor, but peers often see the signs first. A just culture supports both, the right to pause, reset, or challenge, narrowing focus without judgment.

The Role of SOPs and Built-in Buffers

SOPs are not dry rules – they are designed buffers against human limits. They create structure when attention is strained and give people permission to pause and verify. Across aviation, these safeguards appear in many forms: ATC handovers anchored in phraseology, cabin crew cross-checking doors, or maintenance teams using independent sign-offs for high-risk tasks.

Research supports this role. Burian et al. (2005) described SOPs as "decision buffers" that preserve clarity when mental resources are stretched. In practice, they act as early recovery tools, helping crews, controllers, and engineers regain perspective before small lapses escalate. The wider design and psychology of SOPs are explored in *Chapter 9: Application of Procedures: Standard Operating Wisdom.*

Ultimately, decision-making and problem-solving in aviation are never solo acts. They unfold in dynamic systems where clarity, communication, and the courage to question matter as much as technical skill. Every pause, challenge, and shared observation helps transform individual expertise into collective safety. When pilots, cabin crew, controllers, engineers, and dispatchers all commit to noticing and speaking up, they turn the act of deciding into something greater: the continuous practice of keeping each other safe.

TAKEAWAYS

For Pilots and Cabin Crew

- Use short verbal summaries – "here's what we know… here's what's next" – to restore shared awareness when situations become complex or fast-changing.
- Build in structured mini-pauses before acting, especially in non-routine or time-pressured phases of flight or cabin service.
- Raise concerns early, even if they come from intuition or an uneasy feeling. Trust your gut, then validate it with the team.

Reflective Question: *During your last duty, was there a moment when speaking up sooner or pausing briefly might have changed the outcome?*

For Instructors, Examiners, and Leaders

- Reinforce the idea of cognitive resets during Line Oriented Flight Training and CRM sessions. Show how a momentary pause can prevent escalation.
- Coach crews to shift monitoring deliberately, for example, from instruments to outside scan, or from automation to raw data, when workload or automation levels rise.
- In scenario debriefs, highlight the value of "step-back" moments that open up perspective and counter tunneled thinking.

Reflective Question: *In recent training, did you notice a point where learners showed tunnel vision? How could you have guided a reset more effectively?*

For Operational Teams

- Design duty patterns and procedures that allow role alternation or oversight breaks – whether during long-haul cockpit duty, late-night ATC shifts, or extended cabin operations.
- Make monitoring and mitigation training a standard element across all roles, not only for pilots. Cabin crew, engineers, and controllers benefit equally from these skills.
- Support the use of technology, such as adaptive Electronic Flight Bag (EFB) prompts or team-check tools, that help crews and staff maintain active mental scanning under pressure.

Reflective Question: *Looking at your organization's procedures, where could you create more built-in buffers that encourage safe pauses and cross-checks?*

REFERENCES

Aircraft Accident Investigation Board, Pakistan (AAIB Pakistan) (2020) *Aircraft Accident Investigation Report: Pakistan International Airlines Flight 8303, Airbus A320, 22 May 2020.* Available at: https://caapakistan.com.pk/

Bureau d'Enquêtes et d'Analyses pour la sécurité de l'aviation civile (BEA). (2016) *Final report on the accident on 24 March 2015 to the Airbus A320-211, registered D-AIPX, operated by Germanwings, flight GWI18G, near* Prads-Haute-Bléone *(*Alpes-de-Haute-Provence, *France).* Available at: https://bea.aero/uploads/tx_elydbrapports/BEA2015-0125.en-LR.pdf

Burian, B. K., Barshi, I. and Dismukes, K. (2005) *The challenge of aviation emergency and abnormal situations.* NASA TM-2005-213005. Available at: hhttps://ntrs.nasa.gov/api/citations/20060023295/downloads/20060023295.pdf

CAA UK. (2002) *CAP 737: Flight Crew Human Factors Handbook.* Civil Aviation Authority. Available at: https://www.caa.co.uk/publication/download/14984

Caldwell, J. A. and Caldwell, J. L. (2016) *Fatigue in Aviation: A Guide to Staying Awake at the Stick.* Abingdon: Routledge. https://doi.org/10.4324/9781315582030

Dekker, S. (2014) *The Field Guide to Understanding 'Human Error'.* 3rd ed. Boca Raton, FL: CRC Press. https://doi.org/10.1201/9781317031833

Dirección General de Aeronáutica Civil (DGAC, Peru) (2023) *Final Report – Aviation Accident No. 008-2022, LATAM Airlines Flight 2213, Airbus A320-271N (CC-BHB), Lima, 18 November 2022.* Lima: Comisión de Investigación de Accidentes de Aviación (CIAA), Ministry of Transport and Communications, Government of Peru. Available at: https://cdn.www.gob.pe/uploads/document/file/5209692/INFORME%20FINAL%20ACCIDENTE%20DE%20AVIACI%C3%93N%20N%C2%B0%20008-2022.pdf

Directorate General of Civil Aviation (DGCA) (2010) *Report on Accident to Air India Express Flight IX-812 Boeing 737-800, VT-AXV at Mangalore on 22 May 2010.* New Delhi: Ministry of Civil Aviation, Government of India. Available at: https://www.dgca.gov.in/digigov-portal/?baseLocale=en_US?dynamicPage=dynamicPdf/pOKN03WruNcaP9CnNlfjSA%3D%3D&mainAccidentReports/500005/0/viewApplicationDtlsReq

Dreyfus, S. E. and Dreyfus, H. L. (1980) *A Five-Stage Model of the Mental Activities Involved in Directed Skill Acquisition.* Berkeley, CA: University of California. (Operations Research Center). ADA084551. Available at: https://apps.dtic.mil/sti/citations/ADA084551

Dutch Safety Board (2010) *Crashing during approach, Boeing 737-800, Turkish Airlines Flight TK1951, Amsterdam Schiphol Airport, 25 February 2009.* Available at: http://www.onderzoeksraad.nl/docs/rapporten/Rapport_TA_ENG_web.pdf

Endsley, M. R. (1995) 'Toward a theory of situation awareness in dynamic systems', *Human Factors*, 37(1), pp. 32–64. https://doi.org/10.1518/001872095779049543

Federal Aviation Administration (FAA) (1991) *Aeronautical Decision Making for Pilots, Advisory Circular AC 60-22.* Washington, DC: US Department of Transportation. Aviailable at: https://www.faa.gov/documentLibrary/media/Advisory_Circular/AC_60-22.pdf

Federal Aviation Administration (FAA) (2010) *Flight Operational Quality Assurance (FOQA).* Advisory Circular AC 120-82. Available at: https://www.faa.gov/documentLibrary/media/Advisory_Circular/AC_120-82.pdf

Fischhoff, B. (1975)'Hindsight ≠ foresight: The effect of outcome knowledge on judgment under uncertainty', *Journal of Experimental Psychology: Human Perception and Performance*, 1(3), pp. 288–299. https://doi.org/10.1037/0096-1523.1.3.288

Flin, R., O'Connor, P. and Crichton, M. (2008) *Safety at the Sharp End: A Guide to Non-Technical Skills.* Boca Raton, FL: CRC Press. https://doi.org/10.1201/9781315607467

Gillingham, K. K., & Previc, F. H. (1993). *'Spatial orientation in Flight (Tech. Report AL-TR-1993-0022)'.* Brooks Air Force Base, TX: Armstrong Laboratory, Crew Systems Directorate, Crew Technology Division

GPIAAF. (2020) *Serious Incident: Airbus A320-214 G-EZWC, EasyJet Flight 2146, Lisbon Airport, 10 September 2019*. Lisbon. https://skybrary.aero/sites/default/files/bookshelf/5994.pdf

Harris, D. and Li, W.-C. (Eds.) (2015) *Decision Making in Aviation*. Farnham: Ashgate.

Hatano, G. and Inagaki, K. (1986) 'Two courses of expertise', in H. W. Stevenson, H. Azuma, K. Hakuta (Eds.), *Child Development and Education in Japan* (pp. 262–272). New York: W H Freeman/Times Books/Henry Holt & Co.

Hays, R. T., Jacobs, J. W., Prince, C. and Salas, E. (1992) 'Flight simulator training effectiveness: A meta-analysis', *Military Psychology*, 4(2), pp. 63–74. https://doi.org/10.1207/s15327876mp0402_1

Helmreich, R. L. (2000) 'On error management: Lessons from aviation', *BMJ*, 320(7237), pp. 781–785. https://doi.org/10.1136/bmj.320.7237.781

International Civil Aviation Organization (ICAO) (2018) *Safety Management Manual (SMM). 4th ed. Doc 9859. Montreal.* Available at: https://store.icao.int/en/safety-management-manual-doc-9859

Interstate Aviation Committee (IAC/MAK) (2020). *Final Report on the Accident Involving Sukhoi Superjet 100 Aircraft, Registration RA-89098, Operated by Aeroflot – Flight SU1492, Moscow Sheremetyevo Airport, 5 May 2019.* Moscow: Interstate Aviation Committee. Available at:https://mak-iac.org/upload/iblock/b6e/h3kuyl83dvw2c23tetikftjbx7m54civ/report_ra-89098_eng.pdf

Janis, I. L. (1982) *Groupthink: Psychological Studies of Policy Decisions and Fiascoes* 2nd ed. Boston: Houghton Mifflin.

Kahneman, D. (2011) *Thinking, Fast and Slow*. New York: Farrar, Straus and Giroux.

Klein, G. (1998) *Sources of Power: How People Make Decisions*. Cambridge, MA: MIT Press. ISBN 0-262-11227-0.

Klein, G. (2008) 'Naturalistic decision making', *Human Factors*, 50(3), pp. 456–460. https://doi.org/10.1518/001872008X288385

KNKT (2019) *Final aircraft accident investigation report: Lion Air Flight 610, Boeing 737-8 (MAX) PK-LQP, 29 October 2018*. Komite Nasional Keselamatan Transportasi. Available at: https://www.dephub.go.id/post/read/final-report-of-lion-air-flight-610

Komite Nasional Keselamatan Transportasi (KNKT) (2015) *Final Report – Indonesia AirAsia Flight QZ8501 Airbus A320-216, PK-AXC*. Jakarta: KNKT. Available at: https://www.aaiu.ie/sites/default/files/FRA/KNKT%20Indonesia%20Final%20Report%20PK-AXC%20Airbus%20A320-216%20Air%20Asia%20PT%20Indonesia%202015-12-01.pdf

Leach, J. (2004) 'Why people "freeze" in an emergency: Temporal and cognitive constraints on survival responses', *Aviation, Space, and Environmental Medicine*, 75(6), pp. 539–542. Available at: https://www.hptc-pro.com/wp-content/uploads/2014/01/Why-People-Freeze.pdf

Martinussen, M. and Hunter, D. R. (2017) *Aviation Psychology and Human Factors*. 2nd ed. Boca Raton, FL: CRC Press.

NTSB (2020) *Aircraft Accident Report: Loss of Control on Approach, Atlas Air Flight 3591, Boeing 767-375BCF, N1217A, near Anahuac, Texas, February 23, 2019*. NTSB/AAR-20/02. Available at: https://www.ntsb.gov/investigations/AccidentReports/Reports/AAR2002.pdf

O'Connor, P., Flin, R., Fletcher, G., Hemsley, C. and Crichton, M. (2008) 'Methods used to evaluate the effectiveness of CRM training in the UK civil aviation industry', *Human Factors and Aerospace Safety*, 8(4), pp. 117–134.

Orasanu, J. (2002) *Risk Perception: A Critical Element of Aviation Safety*, in *IFAC Proceedings Volumes*, 35(1), pp. 1637–1642. Paper presented at the IFAC World Congress, Barcelona, Spain, July 21–26. https://doi.org/10.3182/20020721-6-ES-1901.01639

Orasanu, J. and Martin, L. (1998) 'Errors in aviation decision making: A factor in accidents and incidents', in C. E. Billings (Ed.), *Aviation Human Factors: Progress in Astronautics and Aeronautics*, Vol. 131. Washington, DC: American Institute of Aeronautics and Astronautics.

Porcelli, A. J. and Delgado, M. R. (2017) 'Stress and decision making: Effects on valuation, learning, and risk-taking', *Current Opinion in Behavioral Sciences*, 14, pp. 33–39. https://doi.org/10.1016/j.cobeha.2016.11.015

Reason, J. (1990) *Human Error*. Cambridge: Cambridge University Press.

Salas, E., Rosen, M. A., Burke, C. S. and Goodwin, G. F. (2009) 'The wisdom of collectives in organizations: An update of the teamwork competencies', in E. Salas, G. F. Goodwin, C. S. Burke (Eds.), *Team Effectiveness in Complex Organizations: Cross-Disciplinary Perspectives and Approaches* (pp. 39–79). Abingdon: Routledge.

Salas, E., Tannenbaum, S. I., Kraiger, K. and Smith-Jentsch, K. A. (2012) 'The science of training and development in organizations: What matters in practice', *Psychological Science in the Public Interest*, 13(2), pp. 74–101. https://doi.org/10.1177/1529100612436661

Salas, E., Wilson, K. A., Burke, C. S. and Wightman, D. C. (2006) 'Does crew resource management training work? An update, extension, and some critical needs', *Human Factors*, 48(2), pp. 392–412. https://doi.org/10.1518/001872006777724444

Stott, J. (2013). 'Orientation and disorientation in aviation', *Extreme Physiology & Medicine*, 2(2), Article 2. https://doi.org/10.1186/2046-7648-2-2

Wickens, C. D. and McCarley, J. S. (2008) *Applied Attention Theory*. Boca Raton, FL: CRC Press. https://doi.org/10.1201/9780429059261

8 Leadership and Teamwork

Coordinating the Invisible

CASE STUDY: INSIDE QANTAS 32 (2010) – A CONTROLLED CHAOS

It began like any other flight. Clear skies over Singapore. Qantas Flight 32 was bound for Sydney with 469 people on board. Captain Richard de Crespigny sat in the left seat for a routine route check, surrounded by an unusually full cockpit: Check Captains David Evans and Harry Wubben, First Officer Matt Hicks, and Second Officer Mark Johnson (ATSB, 2013). On paper, the flight deck was unusually well resourced.

Four minutes after takeoff, everything changed.

Two loud bangs. Not one engine failure – something worse. Debris from engine two tore through the wing, slicing cables, fuel lines, and hydraulics. A fuel tank ignited briefly. Elecronic Centralized Aircraft Monitor (ECAM) lit up with a shocking number of warnings – some overlapping, some contradictory (ATSB, 2013). None of them had ever seen anything like it, not even in the simulator.

"I didn't care if it was one engine or two," de Crespigny said later. "Didn't matter". The crew checked the terrain, set heading hold, and kept them level. The screens glowed red, but the tone stayed calm: "We're at constant altitude, constant heading, constant speed. We are safe." Then they began to work (de Crespigny, 2012).

"This wasn't a technical problem," the captain reflected later. "It was a human performance problem."

No shouting. No chaos. Just quiet coordination. One pilot ran checklists, another monitored systems, someone handled Air Traffic Control (ATC). Leadership flowed naturally – shared, flexible, and situational. It wasn't hierarchy that saved them. It was teamwork and competence (ATSB, 2013; de Crespigny, 2012).

De Crespigny's tone on the flight deck was calm and real. He knew fear spreads faster than smoke. That steady voice became the team's anchor. He also spoke to the cabin: honest, direct, no false comfort. "They needed to know someone was in control," he said (de Crespigny, 2012).

In the cabin, the crew faced their own test. Passengers could see fuel streaming from the wing. Some cried. Others froze. The crew followed procedures but led with humanity – reassuring, grounded, and composed. That steadiness comes from character as much as training (ATSB, 2013).

DOI: 10.1201/9781003515418-8

Fifty checklists followed, yet the team never rushed. They worked methodically until they understood what systems remained. When they finally landed, fuel was leaking, brakes were hot, and one engine still spun. De Crespigny and the flight crew made one last call: wait. The decision reflected continued risk assessment and controlled judgment. Passengers disembarked by stairs an hour later. The rogue engine stopped three hours after touchdown (ATSB, 2013). Nobody was hurt.

That's the headline. But the real story is *how*. No panic. No ego. Just trust and quiet precision. While the captain's leadership was pivotal, the outcome ultimately relied on the entire crew's coordinated management, with each member contributing to the controlled resolution of the crisis. Investigators later traced the failure to a tiny cracked oil pipe – too small to see, too critical to ignore. Flight 32 wasn't luck. It was effective leadership and crew performance under extreme system disruption.

INTRODUCTION

We use the words *leadership* and *teamwork* constantly in aviation, yet their value only shows when things tilt. Leadership on the line isn't rank, and teamwork isn't task-splitting. They are behaviors: tone that slows the room, a timely challenge, a brief check-in that realigns attention (Flin et al., 2008). When the script breaks, what matters is how people communicate, regulate emotion, and share mental workload (Salas et al., 2001)

Qantas 32 made this visible: leadership passed to whoever had the clearest view, while the captain's steadiness held everyone's bandwidth. The same principle scales across roles: cabin crew settling a cabin, engineers framing defects, ATC shaping flow. These behaviors don't appear in a crisis by magic; they are built in everyday briefings, scenario practice, and debriefs in which curiosity replaces blame.

Modern operations are automated, interdependent, and time-tight. Effective teams let leadership flow to expertise, speak plainly, and keep alignment current. This chapter traces how Crew Resource Management (CRM) matured into lived leadership and teamwork, links the behaviors to psychology and brain science, and turns them into trainable practices: variable scenarios, reflective debriefs, inclusive briefings, and mentoring that shapes mindset as well as skill.

In this chapter, we'll look at how leadership and teamwork have evolved in aviation, how they intersect with psychology and training, and how, when lived well, they turn complexity into coordination and uncertainty into calm.

FROM CRM TO LEADERSHIP AND TEAMWORK TRAINING: EVOLVING MODELS

Leadership in aviation has come a long way from the checklist-style authority once taught in early CRM courses. We now know that leadership isn't just about commanding a cockpit – it begins long before the flight and continues in the air. It shapes how people prepare, think, feel, and work together under pressure. This section explores how the DNA of CRM has matured into something deeper and more dynamic – a framework for building real-world leadership and teamwork where it counts most, both before and during the flight.

The Evolution of CRM

CRM laid the foundation for how we understand leadership and teamwork in aviation. Born from tragedy, it reshaped how crews communicate, coordinate, and make decisions under pressure (Wiener et al., 1993). But CRM was only the beginning.

The early courses focused on speaking up, managing hierarchy, and sharing workload. Over time, these principles evolved into something broader – an understanding that leadership and teamwork are living processes, not procedures (Helmreich et al., 1999). Modern crews don't just exchange information; they share awareness, manage emotion, and coordinate cognition under stress.

That shift led to a crucial realization: leadership isn't a title; it's a behavior that moves between people (Flin et al., 2008). The best teams adapt naturally, allowing leadership to flow to whoever has the clearest mental picture at the time.

CRM began the conversation. Leadership and teamwork training keep it alive, shaping aviation into a system where authority and empathy work side by side; an alignment supported by studies showing that empathetic, participative leadership among airline pilots enhances crew satisfaction and collective effort (Rouco et al., 2021).

A Leadership Lens on Modern Training

Leadership in aviation isn't a stand-alone topic anymore; it's woven through everything people do, from simulator briefings and debriefs to multi-crew cooperation and line operations. What used to sit under "CRM training" has become a continuous process that shapes behavior, communication, and judgment across all learning environments.

Modern programs still draw on CRM, but focus on how leadership and teamwork play out under pressure. The goal isn't compliance or checklist thinking; it's helping people adapt, prioritize, and stay composed when plans collapse. Scenario variability and ambiguity, not rehearsed scripts, now anchor development (ICAO, 2026).

Good training accepts that leadership is context-sensitive. A captain may be firm and directive during an immediate failure, then shift to a participative style once stable, and know when to follow (Bastola, 2020). True leadership creates space for others' expertise without ego.

Qantas 32 made this visible: Captain de Crespigny set the tone early, then delegated the flying to First Officer Matt Hicks once the situation stabilized – authority balanced with trust, turning chaos into coordinated action (ATSB, 2013; de Crespigny, 2012).

That judgment grows from feedback, reflection, and role-modelling. When instructors show calm, empathy, and fairness, those qualities cascade into how crews communicate and decide (Edmondson, 1999; 2019). People learn not only from manuals but from the room's tone, mentors' example, and the surrounding safety climate.

This aligns with Safety-II (explored in Chapter 2: *Safety Culture: Beyond Compliance – A Safety Mindset*): prepare for success in uncertainty, not just prevent failure (Hollnagel, 2014). Across the book's positive cases, adaptability and shared decision-making are central. Leadership is practiced long before any crisis in how people prepare, connect, and stay ready. The illustration below depicts leadership as a continuous cycle linking scenario practice, behavioral reflection, and real-world application (Figure 8.1).

The Leader as Mentor

In aviation, the most lasting lessons rarely come from the syllabus, they come from people. The instructor who listens rather than lectures, who models calm after a tense sim, or who admits an error without losing authority teaches leadership by example. These quiet acts of mentorship shape how future captains think under pressure (Kanki et al., 2019).

Leadership as mentorship means teaching others how to reason, not what to remember. It is guiding reflection, curiosity, and composure long after the exercise ends. When instructors debrief not to judge but to understand, they help pilots build the self-awareness and behavioral consistency that underpin safe, adaptive performance (Flin et al., 2003).

As training evolves, mentorship bridges the gap between learning and leadership culture. The next generation of pilots responds not to command but to connection, to leaders who coach, question, and care about how decisions are made (Rouco et al., 2021). This mentoring relationship also strengthens the ***shared mental models*** that underpin effective coordination in flight (see Figure 8.2). In this sense, mentorship becomes the human architecture of flight safety: the transmission of calm, curiosity, and professionalism from one generation to the next.

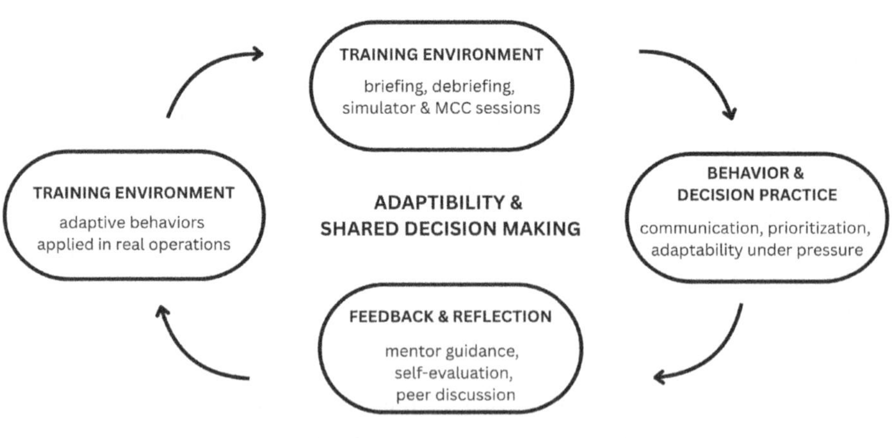

FIGURE 8.1 The leadership development cycle in modern aviation training.

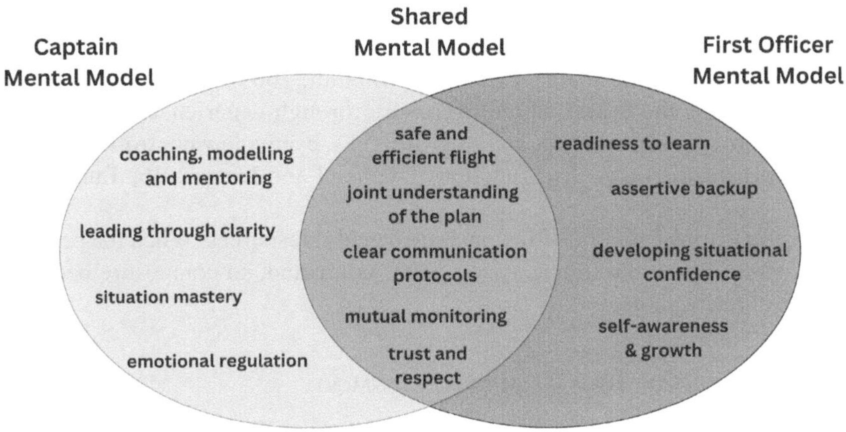

FIGURE 8.2 From individual to shared mental models.

THE NEUROSCIENCE OF LEADERSHIP AND TEAM PERFORMANCE

To lead and work together effectively in modern aviation, it's not enough to manage visible behaviors. Leadership and teamwork are not only social skills, they are grounded in biology. Everything from staying calm to managing workload rests on neural systems constantly tested in high-stakes flight. Understanding how the brain reacts under pressure explains why some leadership habits succeed while others fail.

Leadership as Cognitive Load-Sharing

Flying can challenge the brain's limited processing power. At the center of high-order thinking sits the **prefrontal cortex** – the region just behind the forehead responsible for planning, decision-making, inhibition, and working memory (Miller and Cohen, 2001). It acts like mission control: integrating information from sensory and emotional systems and deciding what deserves attention.

When the workload rises sharply, this control hub can become overloaded. The prefrontal cortex starts dropping tasks, and the more reactive limbic areas (the brain's emotional engines) begin to dominate (Arnsten, 2009). At such moments, leadership is less about authority and more about managing collective cognition.

Effective leaders sense when their own or the team's capacity is thinning. They read silence, tension, and subtle behavioral drift, then delegate, simplify, or pause to restore clarity. This is leadership as **cognitive load-sharing**, which is distributing mental effort rather than hoarding control. Structured handovers and short pauses protect the group's executive bandwidth.

The Qantas 32 crew demonstrated this vividly. Each pilot took defined roles, maintaining joint focus instead of competing for control (ATSB, 2013; de Crespigny,

2012). The captain's composure created thinking space for others. Still, prefrontal performance is fragile: fatigue, circadian disruption, or stress quickly erode working memory and decision quality (Van Dongen et al., 2003). Fortunately, **neuroplasticity,** which is the brain's ability to rewire through experience, also enables recovery. Sleep, aerobic exercise, and micro-breaks, even a few controlled breaths before takeoff, restore executive function (Diamond and Ling, 2016; Tang et al., 2007).

Leadership begins with protecting one's own neural resources. When the prefrontal cortex goes offline, the limbic system takes command, so composure becomes the first act of leadership.

Neurofoundations of Team Calm and Alignment

Operational stress in aviation strikes without warning – a flicker on ECAM, a cabin call, a runway change on final. Before logic can intervene, the limbic system fires. The **amygdala** sounds its alarm, priming the body for action (Arnsten, 2009; LeDoux, 2000). It's a protective reflex, but in the cockpit it can narrow focus, quicken speech, and cloud judgment as prefrontal control temporarily gives way to subcortical dominance (Arnsten, 2009). What distinguishes strong team members under pressure is not the absence of emotion but mastery over its expression.

Resilient leaders don't suppress this response – they regulate it. By breathing steadily and keeping tone measured, they signal safety to their own nervous system and to others. Stress is contagious, but so is calm (Gross, 2015; Thayer and Lane, 2000). Techniques such as tactical breathing, grounding, or naming emotion ("I'm tense") activate prefrontal circuits that inhibit the amygdala's overreaction (Creswell et al., 2007; Lieberman and Eisenberger, 2007; Porges, 2011). Repeated practice gradually reshapes those neural pathways, improving recovery speed and emotional steadiness.

Trust rests on another biological channel, the **oxytocin system**. Oxytocin, a hormone and neurotransmitter produced in the hypothalamus, promotes social bonding, empathy, and cooperative behavior (Carter, 2014; Zak, 2017). In the cockpit it's released through consistent, respectful interaction: a reassuring phrase, an open question, a calm acknowledgment of workload. These micro-signals create psychological safety, the sense that it's safe to speak and contribute.

The Qantas 32 team had never flown together before, yet their predictable tone and transparent dialogue generated instant trust (ATSB, 2013; de Crespigny, 2012). On the other hand, there is also evidence that supports working with familiar crews. They can enjoy **shared mental models,** a cognitive shorthand that speeds coordination and reduces error (Cannon-Bowers et al., 1993), but even ad-hoc teams can align when briefings are explicit and expectations are clear.

Physiology supports or undermines this alignment. Fatigue, caffeine surges, or chronic sleep loss reduce empathy by weakening prefrontal-limbic connectivity (Mileva-Seitz et al., 2012). Supporting rest, nutrition, and recovery becomes a safety imperative, not a luxury. Communication itself is neurobiology in motion: each phrase or pause, either strengthens or strains the brain's cooperative circuits.

Brain Synchronization in High-Functioning Teams

When leadership and teamwork align, a deeper process emerges, which can be called **neural synchrony**. This is the spontaneous alignment of brain activity across individuals engaged in a shared task (Dikker et al., 2017; Stephens et al., 2010). Functional-imaging studies show that during successful cooperation, patterns in auditory, frontal, and parietal regions of team members' brains literally oscillate in rhythm, supporting common attention and timing.

In aviation, this synchrony appears as smooth checklist transitions, timely mutual support, and almost wordless coordination. It's not telepathy, it's physiology tuned by trust, rhythm, and shared focus.

A perfect example came on **Alaska Airlines Flight 1282** in 2024. Shortly after takeoff from Portland, a door plug separated from the fuselage, causing rapid depressurization. The first officer's (FO's) headset was torn away, and the intense cabin noise made cockpit communication extremely difficult. As the FO recalled, the cabin noise made it hard for the first officer to hear the captain, and the right-seat checklist wasn't immediately accessible. Without confusion, the captain handed over the quick reference handbook and the first officer read the procedure aloud. Communication between cockpit and cabin was also difficult in both directions (each side uncertain of the other's status), but a shared mental picture and synchrony held: priorities, next steps, and task division. Despite degraded communications, alignment and tempo stayed intact (Miller, 2024; NTSB, 2025).

Synchrony is delicate. Fatigue, emotional tension, or uncertainty can break alignment, visible in mistimed handovers or delayed actions. **Physiological stability**, so adequate rest, glucose balance, and calm breathing – preserves the link; deliberate verbal pacing restores it. When a leader slows speech or names pressure, the crew's physiology often follows.

When brains sync, crews function as one. They adapt faster, communicate more precisely, and share workload instinctively. Neural synchrony represents the biological end-state of effective leadership and teamwork, a human network thinking as a single cognitive system.

Table 8.1 summarizes the key neural systems and how they translate into operational behavior, closing the loop between biology and cockpit performance.

HOW WE SHOULD MEASURE LEADERSHIP AND TEAMWORK?

Leadership and teamwork may seem intangible, but in aviation they can, and must, be measured. We do it every day through what we see and hear: tone, timing, cooperation, and decision-making under pressure. The Competency-Based Training and Assessment (CBTA) model gives us a shared language for this. It translates abstract

TABLE 8.1
Neurocognitive Skills and Leadership Behaviors

Neurocognitive Process	Associated Leadership Behavior	Impact on Team Performance	Influencing Factors
Executive Function (Prefrontal Cortex)	Prioritizing, delegating, decision flow	Maintains mental clarity, reduces overload, supports safe outcomes	Sleep quality, chronic stress, task rehearsal, and workload design
Emotional Regulation (Amygdala)	Remaining calm under pressure	Stabilizes team response, prevents emotional contagion	Breathing techniques, resilience routines, recent recovery time
Oxytocin Pathways	Building trust, encouraging input	Enhances psychological safety and openness to correction	Communication quality, familiarity, social connectedness
Neural Synchrony	Shared mental models, synchronized timing	Enables faster coordination and adaptive response to dynamic events	Mutual respect, sleep, fatigue, focused communication

human qualities into observable, assessable actions. Still, what we observe is only the surface. Beneath it lie emotional awareness, mindset, and cultural context, all of which determine how those behaviors emerge.

Observable Behaviors as a Solid Foundation

As discussed earlier, different organizations define leadership and teamwork in slightly different ways. Still, all CBTA frameworks (whether from ICAO, International Civil Aviation Organization; EASA, European Union Aviation Safety Agency; or local training systems), share one principle: these competencies are expressed through **observable behaviors (OBs)**. These indicators may not capture every nuance, but they offer a reliable starting point for training and assessment.

Under **EASA's KSA 100** framework, student pilots are already observed for leadership and teamwork early in training. It's not just about how they fly, but how they lead themselves, collaborate, and communicate within shared tasks. These skills mature over time with experience, exposure, and reflection (EASA, 2018). For instructors and assessors, OBs provide clarity and consistency. They make evaluation objective, even in fast-paced simulator sessions. But they're just the surface layer. Underneath are traits, habits, and emotional patterns that define how people lead under pressure. Those aren't always easy to grade – yet they matter just as much.

The next sections explore those deeper layers: **emotional intelligence, attitude, and culture**. They may not sit neatly on an assessment sheet, but they shape every interaction in training and operation alike.

Emotional Intelligence (Competence) in the Flight Deck

Emotional intelligence doesn't usually appear on checklists – but you can feel when it's there. It's in the pilot who notices a colleague's rising stress and softens their tone instead of matching it. It's the quiet ability to pause before reacting defensively. It's recognizing that silence can mean focus, not resistance.

While the idea of **emotional intelligence (EI)** gained wide attention in the 1990s (Goleman, 1995), research since has questioned how precisely it can be defined or measured. Some psychologists argue that EI overlaps strongly with well-validated traits such as agreeableness, openness, or conscientiousness from the Big Five model (Joseph and Newman, 2010; Landy, 2005). In aviation, however, we're less concerned with the label and more with the **capability** – the ability to recognize and regulate emotion, empathize with others, and sustain effective communication under pressure.

Many now use the term **emotional competence** to emphasize observable behavior rather than theoretical scoring (Matthews et al., 2019). In the cockpit, it becomes tangible: how we listen when workload is high, how we give feedback without judgment, and how we create room for others to contribute. However, the term *emotional intelligence* remains more familiar to most aviation professionals and continues to serve as a practical shorthand for these same interpersonal capabilities.

Teams with strong emotional competence don't just manage tasks – they manage tension. They recover faster from missteps, preserve psychological safety, and maintain focus even when decisions are tough or time is tight. We may not always assess it formally, but we constantly experience its effects.

Attitude: The Unseen Variable

Attitude isn't about optimism; it's about *orientation* – how someone approaches feedback, authority, and uncertainty. It shapes how people react when challenged or corrected.

Some pilots display strong technical skills yet resist input or shift blame when things go wrong. Others show quiet confidence, humility, and openness to learning. The difference is attitude, and it affects both performance and safety.

Attitude also interacts with **bias**. Subtle judgments around age, rank, gender, or culture can erode teamwork if left unchecked. Skilled leaders stay aware of these dynamics and actively counter them through fairness and respect (Helmreich and Merritt, 1999).

Although attitude isn't a formally rated CBTA competency, it can be assessed through **scenario-based exercises** and **structured professional interviews**. In selection and training contexts, these methods reveal how candidates handle ambiguity, feedback, and interpersonal tension. Well-designed behavioral interviews probe real examples: "Tell me about a time you disagreed with a captain" or "Describe how you adapted to unexpected feedback," which provide insight into openness, accountability, and learning orientation (Bartram, 2005; Hunter, 2003). Similarly, **scenario-based assessments** or CRM exercises expose underlying attitudes to teamwork, authority gradients, and safety culture in observable ways (Flin et al., 2003).

Culture and Context

Culture doesn't just mean nationality. It includes **organizational norms**, **generational beliefs**, and **operational subcultures** – all of which affect how people give feedback, ask for help, or assert authority.

A crew member from a hierarchical background may hesitate to challenge decisions. A junior pilot who's been trained in an open, coaching environment might expect dialogue, where others expect deference. Neither is wrong – but the clash can impact coordination.

Understanding cultural framing allows leaders to adapt without losing clarity (Helmreich and Merritt, 1999).

This is where **diversity, equity, and inclusion** (DEI) move from corporate posters to real-world dynamics. Understanding these cultural frames, and recognising how inclusion shapes trust, helps leaders move diversity from policy to practice (Roberson, 2006).

FROM CONTROL TO COORDINATION

Every effective crew learns that leadership is not about holding control, it's about maintaining connection. On the flight deck, authority may start with the captain, but coordination depends on how people think together. As operations become more complex and automated, leadership is less about directing tasks and more about aligning minds. What keeps a team safe is not one person's command, but a shared cognitive rhythm, continuous monitoring, recalibration, and trust in how others think.

The following sections explore this shift from control to coordination, showing how leaders use cognition, reflection, and attention management to keep teams adaptive under pressure.

COGNITIVE ADAPTABILITY AND UPDATING

The human brain is designed to adjust. When new information contradicts expectation, the **anterior cingulate cortex** and **prefrontal regions** register a prediction error and trigger re-evaluation (Friston, 2010; Joiner et al., 2017). In aviation, that moment might appear as a quiet pause, a second look at an instrument, or a simple "Does that make sense?" High-performing teams act on those micro-signals instead of ignoring them, keeping their collective picture of the situation current and aligned.

A clear modern example came from **United Airlines Flight 328** in February 2021. Minutes after take-off from Denver, the Boeing 777's right engine suffered an uncontained failure, scattering debris across nearby suburbs. Captain Chris Dann and First Officer William McGinnis stabilized the aircraft, declared an emergency, and prepared to return. Despite the shock and workload, communication stayed calm and structured. The captain coordinated with ATC while the first officer ran checklists and confirmed actions aloud. Cabin crew, briefed early and clearly, managed passengers who could see the damaged engine.

What held the operation together wasn't just professionalism; it was real-time cognitive updating. Both pilots continuously questioned assumptions, integrated new data, and recalibrated their shared understanding of the flight. That quiet discipline allowed seamless division of tasks and composure under pressure. The aircraft landed safely back in Denver with no injuries (NTSB, 2024).

Metacognition and Collaborative Reasoning

In complex operations, knowing *what* to think is rarely enough; crews must also remain aware of *how* they are thinking. **Metacognition**, so the ability to reflect on one's own reasoning in real time, allows teams to monitor assumptions, detect bias, and recognise when confidence begins to exceed available evidence (Flavell, 1979). In the cockpit, this is not a purely individual skill. When reasoning is verbalised, it becomes shared, testable, and open to challenge. Leadership then shifts from command to dialogue, creating space for correction before errors become embedded.

LOT Polish Airlines Flight 16 from 2011 provides a revealing example of collaborative reasoning under sustained workload. En route from Newark to Warsaw, the Boeing 767 experienced a failure of the normal landing gear extension system. Over the course of the transatlantic flight, the crew managed fuel dumping, coordinated emergency services, and prepared the cabin, while maintaining calm, structured, and respectful communication. Intentions and actions were discussed openly, supporting a shared understanding throughout the event.

The aircraft ultimately conducted a controlled gear-up landing without loss of life. The subsequent investigation confirmed that the aircraft remained largely functional. A circuit breaker supplying the alternate hydraulic system had tripped; resetting it would likely have enabled normal gear extension. The investigation did not identify deficiencies in crew competence or professionalism, noting instead disciplined teamwork under prolonged operational demand (PKBWL, 2017).

Used here, is not to discuss the performance, but to illustrate how attention and reasoning can become progressively organized around a dominant operational interpretation during extended abnormal situations. Once a situation appears technically understood and procedurally managed, teams may naturally focus on consequence management rather than re-opening earlier assumptions. The case highlights that team-level metacognition is not only about open communication, but about sustaining active doubt – the continued willingness to ask what might still be missing, even when the situation feels under control.

The example highlights that metacognition at the team level is not only about speaking openly, but about sustaining active doubt – the willingness to revisit what might still be wrong, even when the situation appears fully understood.

Attention Shifting and Leadership Fluidity

Modern leadership demands flexibility: knowing when to direct, when to share, and when to follow. The parietal and frontal attention networks enable this by switching focus and redistributing cognitive load (Corbetta and Shulman, 2002). In practice, a captain may move from a firm, task-focused style during a system failure to a participative tone once the aircraft is stable. A first officer may take initiative when workload peaks. These transitions maintain team tempo without confusion.

Leadership in this sense is not a chain of command but a chain of cognition – a continuous exchange of focus and control that keeps the group thinking as one. When attention is shared deliberately, leadership flows to expertise, not hierarchy. That is the essence of cognitive coordination: distributed thinking guided by trust, clarity, and calm communication.

THE FUTURE OF LEADERSHIP AND TEAMWORK: EVOLVING MINDS, EMERGING GENERATIONS

Leadership and teamwork in aviation are changing shape. Automation, connectivity, and new generations entering the flight deck are redefining what "authority" and "team" mean. The strongest crews of tomorrow won't just manage technology, they'll manage cognition, emotion, and diversity of thought.

Modern Leadership Culture

Today's leadership is behavioral, not positional. Authority still matters, but influence travels through credibility, empathy, and composure. Instructors, line captains, and training leaders are now expected to coach, not command, to create thinking environments where others can grow.

Leadership culture becomes visible in everyday micro-behaviors: how people debrief, how disagreement is handled, how inclusion feels. When those interactions model fairness and curiosity, leadership becomes a shared norm rather than a title. After Qantas Flight 32, Qantas CEO Alan Joyce's decision to ground the A380 fleet was a symbolic act of trust in the crew and the system, a demonstration that integrity and accountability define culture more than image (ATSB, 2013; de Crespigny, 2012).

Distributed Teamwork in Complex Systems

Modern teamwork goes far beyond the cockpit. Today, safety and performance depend on fluid coordination across multiple roles, time zones, and digital interfaces. Flight operations, maintenance, dispatch, and cabin service are no longer separate silos; they are part of one extended cognitive system, sharing situational awareness in real time.

As automation becomes more advanced, the need for human coordination across roles only grows stronger. Technology can connect systems instantly, but people still need to connect **meaningfully**. The more tasks automation absorbs, the more essential it becomes to align human reasoning, so that every role understands not only *what* others are doing, but *why*. Research across collaborative domains shows that shared cognition and cross-boundary communication are key predictors of resilient team performance (Salas et al., 2017).

That mindset must start early. Coordination across disciplines should be taught from the beginning of training, not as an afterthought, but as a natural part of how aviation professionals think and cooperate (EASA, 2022). Too often, the physical environment divides: pilots behind cockpit doors, engineers in hangars, cabin crew in separate briefings. True safety culture requires bringing those worlds closer, helping each group appreciate how others think, decide, and communicate.

A matching illustration came from **Air France Flight 066** in 2017. When the A380 suffered an uncontained engine failure over Greenland, the crew managed the situation through a structured division of tasks, continuous monitoring across roles, and clear, timely communication with ATC, dispatch, cabin crew, and passengers. Monitoring was not confined to the flight deck but distributed across the wider system, with information actively shared, verified, and updated as the situation evolved. The event demonstrated how effective monitoring culture supports decision-making under uncertainty, allowing the crew to maintain control, situational awareness, and coordination despite significant technical disruption and system degradation (BEA, 2020).

The next stage of aviation teamwork lies in teaching cooperation as a shared professional identity, not a job-specific skill. Leadership must move laterally, not just vertically, across systems and roles as well as within them. The future belongs to teams that think together, regardless of where they sit on the aircraft or in the operation.

New Generations, New Dynamics

Every generation redefines what leadership feels like. The newest wave of aviators enters a world where communication is instant, information is shared, and hierarchy matters less than collaboration. They have grown up working in networks, not ladders, and expect feedback, transparency, and purpose from the start (Deloitte, 2024).

This digital fluency can strengthen teamwork across roles. Younger pilots, cabin crew, and engineers are already comfortable collaborating through connected platforms and real-time data. They intuitively treat information as something to be exchanged, not guarded – a mindset that supports the distributed teamwork modern aviation increasingly depends on (Salas et al., 2017). When channeled through structured CRM and CBTA frameworks, this openness becomes an operational advantage: faster learning, clearer coordination, and more resilient communication under stress.

But generational diversity also challenges traditional norms. New entrants may question command style, authority distance, or communication tone, sometimes interpreted by older colleagues as lack of respect rather than difference in framing. Leadership training must therefore include **inter-generational awareness**: understanding how values, feedback expectations, and digital habits shape behavior on the line (Rouco et al., 2021). When leaders adapt their communication style, offering reasoning, not just instruction, they bridge this gap and turn diversity into collective strength.

Cross-role cooperation will increasingly depend on this blend of **digital confidence and emotional intelligence**. As automation takes over mechanical routine, the human edge becomes social and cognitive: how quickly people can align thinking, update shared awareness, and sustain trust across backgrounds and generations. The next frontier of leadership in aviation will be teaching crews to integrate, not only systems and roles, but also perspectives, identities, and ways of thinking.

TAKEAWAYS

For Pilots and Cabin Crew

- Make feedback a habit – use briefings and debriefings as spaces to reinforce leadership and teamwork behaviors.
- Recognize the power of tone and body language in shaping team calm and trust.
- Seek and offer mentorship across experience levels – it helps build shared confidence and resilience.

Reflective Question: *What small actions do I take daily that contribute to (or undermine) psychological safety on my team?*

For Instructors, Examiners, and Leaders

- Model inclusive, emotionally intelligent leadership – especially under pressure.
- Integrate teamwork observations into technical assessments and feedback.
- Actively shape cross-functional mental models through shared training and dialogue.

Reflective Question*: How do I mentor others – not just in skill, but in mindset and collaboration?*

For Operational Teams

- Structure leadership and team Key Performnce Indicators (KPIs) into both training and operational performance metrics.
- Ensure top-down alignment by training managers and directors in collaborative leadership models.
- Invest in long-term coaching and structured mentoring programs for new leaders.

Reflective Question*: How does our organizational culture reward and support the type of leadership and teamwork we need in the future?*

REFERENCES

Arnsten, A. F. T. (2009) 'Stress signalling pathways that impair prefrontal cortex structure and function', *Nature Reviews Neuroscience*, 10(6), pp. 410–422. https://doi.org/10.1038/nrn2648

Australian Transport Safety Bureau (2013) *In-Flight Uncontained Engine Failure – Airbus A380-842, VH-OQA, Near Batam Island, Indonesia, 4 November 2010 (AO-2010-089).* Canberra: ATSB. Available at: https://www.atsb.gov.au/publications/investigation_reports/2010/aair/ao-2010-089/

Bartram, D. (2005) 'The great eight competencies: A criterion-centric approach to validation', *Journal of Applied Psychology*, 90(6), pp. 1185–1203. https://doi.org/10.1037/0021-9010.90.6.1185

Bastola, D. P. (2020) 'The relationship between leadership styles and aviation safety: A study of the aviation industry', *Journal of Air Transport Studies*, 11(1), pp. 71–102. https://doi.org/10.38008/jats.v11i1.155

Bureau d'Enquêtes et d'Analyses pour la Sécurité de l'Aviation Civile (BEA) (2020) *Accident Report: Air France Flight 066, Airbus A380-861, Engine No. 4 Failure, Over Greenland, September 30, 2017.* Paris: BEA.

Cannon-Bowers, J. A., Salas, E. and Converse, S. (1993) 'Shared mental models in expert team decision making', in N. J. Castellan, Jr. (Ed.), *Individual and Group Decision Making: Current Issues* (pp. 221–246). Mahwah, NJ: Lawrence Erlbaum Associates, Inc.

Carter, C. S. (2014) 'Oxytocin pathways and the evolution of human behavior', *Annual Review of Psychology*, 65, pp. 17–39. https://doi.org/10.1146/annurev-psych-010213-115110

Corbetta, M. and Shulman, G. L. (2002) 'Control of goal-directed and stimulus-driven attention in the brain', *Nature Reviews Neuroscience*, 3(3), pp. 201–215. https://doi.org/10.1038/nrn755

Creswell, J. D., Way, B. M., Eisenberger, N. I. and Lieberman, M. D. (2007) 'Neural correlates of dispositional mindfulness during affect labelling', *Psychosomatic Medicine*, 69(6), pp. 560–565. https://doi.org/10.1097/PSY.0b013e3180f6171f

de Crespigny, R. (2012) *QF32*. Sydney: Pan Macmillan Australia. ISBN 978-1-74261-117-4. (Memoir and primary first-hand account by Captain Richard de Crespigny.)

Deloitte (2024) *Deloitte Global 2024 Gen Z and Millennial Survey*. London: Deloitte. Available at: https://www.deloitte.com/global/en/issues/work/deloitte-2024-genz-millennial-survey.html

Diamond, A. and Ling, D. S. (2016) 'Conclusions about interventions, programs, and approaches for improving executive functions that appear justified and those that, despite much hype, do not', *Developmental Cognitive Neuroscience*, 18, 34–48. https://doi.org/10.1016/j.dcn.2015.11.005

Dikker, S., Wan, L., Davidesco, I., et al. (2017) 'Brain-to-brain synchrony tracks real-world dynamic group interactions in the classroom', *Current Biology*, 27(9), pp. 1375–1380. https://doi.org/10.1016/j.cub.2017.04.002

Edmondson, A. (1999) 'Psychological safety and learning behavior in work teams', *Administrative Science Quarterly*, 44(2), 350–383. https://doi.org/10.2307/2666999

Edmondson, A. (2019) *The Fearless Organization: Creating Psychological Safety in the Workplace for Learning, Innovation, and Growth*. Hoboken, NJ: Wiley.

European Union Aviation Safety Agency (EASA) (2018) *ED Decision 2018/001/R – Part-FCL; Area 100 KSA (KSA 100).* Available at: https://www.easa.europa.eu/en/document-library/agency-decisions/ed-decision-2018001r

European Union Aviation Safety Agency (EASA) (2022) *Human Factors in Aviation Maintenance and Engineering: Training for the Future Workforce*. Cologne: EASA.

Flavell, J.H. (1979). 'Metacognition and cognitive monitoring: A new area of cognitive–developmental inquiry', *American Psychologist*, 34(10), 906–911. https://doi.org/10.1037/0003-066X.34.10.906

Flin, R., Martin, L., Goeters, K. M., Hörmann, H. J., Amalberti, R., Valot, C. and Nijhuis, H. (2003) 'Development of the NOTECHS (non-technical skills) system for assessing pilots' CRM skills', *Human Factors and Aerospace Safety*, 3(2), pp. 97–119.

Flin, R., O'Connor, P. and Crichton, M. (2008) *Safety at the Sharp End: A Guide to non-Technical Skills*. Aldershot: Ashgate.

Friston, K. (2010) 'The free-energy principle: A unified brain theory?', *Nature Reviews Neuroscience*, 11(2), pp. 127–138. https://doi.org/10.1038/nrn2787

Goleman, D. (1995) *Emotional Intelligence: Why It Can Matter More than IQ*. New York: Bantam Books.

Gross, J. J. (2015) 'Emotion regulation: Current status and future prospects', *Psychological Inquiry*, 26(1), pp. 1–26. https://doi.org/10.1080/1047840X.2014.940781

Helmreich, R. L. and Merritt, A. C. (1999) *Culture at Work in Aviation and Medicine: National, Organizational and Professional Influences*. Aldershot: Ashgate.

Helmreich, R. L., Merritt, A. C. and Wilhelm, J. A. (1999) 'The evolution of crew resource management training in commercial aviation', *International Journal of Aviation Psychology*, 9(1), pp. 19–32. https://doi.org/10.1207/s15327108ijap0901_2

Hollnagel, E. (2014) *Safety-I and Safety-II: The Past and Future of Safety Management*. Farnham: Ashgate.

Hunter, D. R. (2003) 'Measuring general aviation pilot judgment using a situational judgment technique', *The International Journal of Aviation Psychology*, 13(4), pp. 373–386. https://doi.org/10.1207/S15327108IJAP1304_03

International Civil Aviation Organization (ICAO) (2026) *Manual of Evidence-Based Training – Volume I: Aeroplane Pilots (Doc 9995-1)*, 2nd edn (Advance version). Montreal: ICAO. Available at: https://store.icao.int/en/manual-of-evidence-based-training-volume-i-aeroplane-pilots-doc-9995-

Joiner, J., Piva, M., Turrin, C. and Chang, S. W. C. (2017) 'Social learning through prediction error in the brain', *npj Science of Learning*, 2, Article 8. https://doi.org/10.1038/s41539-017-0009-2

Joseph, D. L. and Newman, D. A. (2010) 'Emotional intelligence: An integrative meta-analysis and cascading model', *Journal of Applied Psychology*, 95(1), pp. 54–78. https://doi.org/10.1037/a0017286

Kanki, B. G., Helmreich, R. L. and Anca, J. (2019) *Crew Resource Management*. 3rd ed. London: Academic Press.

Landy, F. J. (2005) 'Some historical and scientific issues related to research on emotional intelligence', *Journal of Organizational Behavior*, 26(4), pp. 411–424. https://doi.org/10.1002/job.317

LeDoux, J. E. (2000) 'Emotion circuits in the brain', *Annual Review of Neuroscience*, 23, pp. 155–184. https://doi.org/10.1146/annurev.neuro.23.1.155

Lieberman, M. D., Eisenberger, N. I., et al. (2007) 'Putting feelings into words: Affect labeling disrupts amygdala activity', *Psychological Science*, 18(5), pp. 421–428.

Matthews, G., Zeidner, M. and Roberts, R. D. (2019) *Emotional Intelligence: A Framework for Assessing Emotional Competence*. 2nd ed. New York: Springer.

Mileva-Seitz, V. R., Bakermans-Kranenburg, M. J. and van IJzendoorn, M. H. (2012) 'Sleep and bonding: The role of oxytocin in parental empathy', *Sleep Medicine Reviews*, 16(4), pp. 437–448. https://doi.org/10.1016/j.smrv.2011.09.002

Miller, E. K. and Cohen, J. D. (2001) 'An integrative theory of prefrontal cortex function', *Annual Review of Neuroscience*, 24, pp. 167–202. https://doi.org/10.1146/annurev.neuro.24.1.167

Miller, J. (2024) *Alaska Airlines Flight 1282 First Officer interview*. YouTube, January 9. Available at: https://www.youtube.com/watch?v=5pnLVjLhcKc&t=1333s

National Transportation Safety Board (2025) *In-Flight Separation of Left Mid Exit Door Plug, Alaska Airlines Flight 1282, Boeing 737-9, N704AL, Portland, Oregon, January 5, 2024. Aviation Investigation Report AIR-25-04*, adopted June 24, 2025. Washington, DC: NTSB. Available at: https://www.ntsb.gov/investigations/AccidentReports/Reports/AIR2504.pdf

National Transportation Safety Board (NTSB) (2024) *Accident Report: United Airlines Flight 328, Boeing 777-222, N772UA, Engine Failure and In-Flight Fire, Denver, Colorado, February 20, 2021 (DCA21FA085)*. Washington, DC: NTSB. Available at: https://www.ntsb.gov/investigations/Pages/DCA21FA085.aspx

Państwowa Komisja Badania Wypadków Lotniczych (PKBWL) (2017) *Final Report: Boeing 767-300ER, SP-LPC, LOT Polish Airlines Flight LO16, Warsaw Chopin Airport (EPWA), November 1, 2011 (Event No. 2011/1400). Warsaw: PKBWL*. Available at: https://pkbwl.gov.pl/wp-content/uploads/2023/11/2011_1400_RK_ENG.pdf (https://pkbwl.gov.pl/raporty/2011-1400/)

Porges, S. W. (2011) *The Polyvagal Theory: Neurophysiological Foundations of Emotions, Attachment, Communication, and Self-Regulation*. New York: W. W. Norton & Company.

Roberson, Q. M. (2006) 'Disentangling the meanings of diversity and inclusion in organizations', *Group & Organization Management*, 31(2), pp. 212–236. https://doi.org/10.1177/1059601104273064

Rouco, C., Soares, M. E., Mendes, E. and Jamal, S. (2021) 'Leadership style and communication style of airline pilots: Perceived associations with crew members' satisfaction and effort', *International Journal of Aviation, Aeronautics, and Aerospace*, 8(4), Article 2. DOI: https://doi.org/10.15394/ijaaa.2021.1627

Salas, E., Burke, C. S., Bowers, C. A. and Wilson, K. A. (2001) 'Team training in the skies: Does crew resource management (CRM) training work?', *Human Factors*, 43(4), pp. 641–674. https://doi.org/10.1518/001872001775870386

Salas, E., Rico, R. and Passmore, J. (2017) *The Wiley Blackwell Handbook of the Psychology of Team Working and Collaborative Processes*. Oxford: Wiley-Blackwell.

Stephens, G. J., Silbert, L. J. and Hasson, U. (2010) 'Speaker–listener neural coupling underlies successful communication', *Proceedings of the National Academy of Sciences of the United States of America*, 107(32), pp. 14425–14430. https://doi.org/10.1073/pnas.1008662107

Tang, Y.-Y., Ma, Y., Wang, J., Fan, Y. and Posner, M. I. (2007) 'Short-term meditation training improves attention and self-regulation', *Proceedings of the National Academy of Sciences*, 104(43), pp. 17152–17156. https://doi.org/10.1073/pnas.0707678104

Thayer, J. F. and Lane, R. D. (2000) 'A model of neurovisceral integration in emotion regulation and dysregulation', *Journal of Affective Disorders*, 61(3), pp. 201–216. https://doi.org/10.1016/S0165-0327(00)00338-4

Van Dongen, H. P. A., Maislin, G., Mullington, J. M. and Dinges, D. F. (2003) 'The cumulative cost of additional wakefulness', *Sleep*, 26(2), pp. 117–126. https://doi.org/10.1093/sleep/26.2.117

Wiener, E. L., Kanki, B. G. and Helmreich, R. L. (1993) *Cockpit Resource Management*. San Diego, CA: Academic Press.

Zak, P. J. (2017) *The Moral Molecule: The Source of Love and Prosperity*. New York: Dutton.

9 Application of Procedures

Standard Operating Wisdom

CASE STUDY: CATHAY PACIFIC FLIGHT 780 (2010) – WHEN PROCEDURES AREN'T ENOUGH

April 13, 2010, began like any other day for Captain Malcolm Waters and First Officer David Hayhoe. Their Airbus A330, operating as Cathay Pacific Flight 780, departed Surabaya, Indonesia, bound for Hong Kong with 309 passengers and 13 crew. Clear skies, smooth cruise – nothing unusual. The kind of flight that quietly reinforces how reliable modern aviation has become.

Then, mid-cruise, the first alert appeared. Engine number 2 had developed an electronic control fault. It could still run, but now required manual management. Not ideal, but the crew knew the drill. The Standard Operating Procedures (SOPs) covered this scenario neatly. They followed the checklist, stabilized the system, and prepared for descent.

Moments later, engine number 1 began to fail, and this time the situation fell outside anything the manuals anticipated. The left engine was stuck at near full power, the right at idle, one side pushing forward, the other barely producing thrust. The flight management system couldn't reconcile the inputs. Warnings flashed, automation disengaged, and the aircraft began to behave unpredictably. There was no checklist for this.

The SOPs, designed for single-engine operations or dual flameouts, offered no guidance for an asymmetric thrust lock. The crew used what they could, shutting down distractions, prioritizing systems, and managing workload. Captain Waters flew manually while First Officer Hayhoe handled communications and calculations. With throttles frozen, speed brakes and landing gear became their only tools for energy control. They chose Hong Kong's long Runway 07L and prepared for a high-speed approach.

Touchdown came at approximately 231 knots (~428 km/h). The A330 slammed into the runway, asymmetrical thrust pulling it sideways. Brakes glowed, reverse thrust on one engine fought the other still pushing forward, and hydraulic systems strained. Yet somehow, the aircraft stopped – near the end of the runway. Fifty-seven passengers and six cabin crew were injured during the evacuation, however mostly minor (CAD, 2013).

Investigators later found that fuel from Surabaya had been contaminated with super-absorbent polymer particles, jamming both engines' fuel-metering

DOI: 10.1201/9781003515418-9

valves (CAD, 2013). The crew had done nothing wrong; in fact, they received the International Federation of Air Line Pilots' Associations (IFALPA) Polaris Award for exceptional airmanship (ALPA Japan, 2014).

Their achievement lay not only in technical skill but in mindset. They followed SOPs as far as possible, then transcended them with calm, analytical improvisation. True procedural wisdom, their flight proved, lies not in memorizing steps but in grasping their intent, the mastery that bridges the gap when the book runs out of pages.

INTRODUCTION

SOPs are one of the greatest human inventions in aviation safety. They exist to standardize complex operations, minimize uncertainty, support coordination, and reduce the mental load in high-stakes environments (Degani and Wiener, 1993; ICAO, 1998). They are checklists, scripts, flow patterns, and decision trees all rolled into one. For many, they are a kind of cognitive compass – a structure that keeps focus steady when pressure rises (Wiegmann and Shappell, 2017).

But SOPs are not infallible. They do not think, they do not adapt, and they cannot replace judgment. When used well, SOPs are a scaffold for thinking. When we use them blindly, they risk becoming a barrier to insight (Hollnagel, 2009).

Although this chapter might seem to focus mainly on pilots, its principles extend far beyond the cockpit. Cabin crew, engineers, air traffic controllers, and ground operations staff all depend on procedures to manage workload, communicate effectively, and maintain consistency under time pressure (Flin et al., 2008). The human factors behind designing, following, and interpreting SOPs – clarity, logic, brevity, and adaptability – are universal. Whether it's a maintenance task card, a cabin safety briefing, or a flight deck checklist, the psychology of procedure remains the same: structure supports safety only when it works with the human mind (Dekker, 2017; Reason, 2016).

The following pages explore this psychology: how SOPs help the brain under pressure, how they interact with assertiveness and culture, and where the line lies between procedural discipline and procedural rigidity. They also look at how SOPs are trained, designed, misunderstood, and improved, and why the real skill lies not just in compliance, but in procedural wisdom.

WHY THE BRAIN LOVES PROCEDURES: THE NEUROPSYCHOLOGY OF SOPs

Even the most experienced operators know that when things go wrong, instinct alone is not enough. In the chaos of alarms, distractions, or competing demands, the brain begins to misfire. Thoughts scatter, perception narrows, and what once felt like skill can suddenly feel like confusion (Kahneman, 2011; Wickens, 2008). In those moments, structure saves lives. SOPs exist not to limit intelligence but to support it – to give the human mind something stable to hold on to when its own circuitry begins to tremble (Flin et al., 2008; Reason, 2016).

For pilots, this may mean the comfort of a well-worn memory flow or an emergency checklist. For an air traffic controller, it's the phraseology and separation protocols that shape calm communication under pressure. For a maintenance engineer, it's the torque value and sign-off routine that ensures precision even after a long shift. Across every role, SOPs are the nervous system of aviation – keeping order when the human brain starts to drift toward chaos (Dekker, 2017).

SOPs as Externalized Memory

In the moment of an emergency, when heart rate spikes and attention narrows like a tunnel, people don't reach for creativity – they reach for structure. It is in these moments that SOPs transform from written rules into lifelines.

From a neuropsychological perspective, SOPs act as **externalized memory**. They offload information from short-term working memory, which can hold only four to seven discrete items at a time (Cowan, 2010). Under pressure, this capacity shrinks even further as **cortisol** and **noradrenaline** flood the system, weakening the prefrontal cortex and reducing flexibility (Arnsten, 2009).

By providing clear, sequential actions, SOPs act as an outline for cognition – stabilizing behavior when reasoning begins to fragment. This mechanism is well supported by research into human performance under stress, which shows that written or visual cues can restore operational accuracy, even when mental capacity declines (Hockey, 2013). NASA's human factors studies have also shown that in high-workload cockpit simulations, checklists serve as external "working memory buffers," preventing key omissions during startle or overload (Casner and Schooler, 2015).

Predictability Reduces Load

Procedures create predictability – something the brain craves. Predictable patterns reduce mental workload and restore a sense of control. Once routine actions become automatic, cognitive energy can be redirected to monitoring, communication, or detecting subtle anomalies.

This predictability doesn't just aid the individual – it synchronizes the team. Shared procedural language allows pilots, controllers, and cabin crew to anticipate one another's actions with minimal verbal load. The International Civil Aviation Organization's (ICAO's) Human Factors Training Manual (Doc 9683) highlights that standardized routines reduce ambiguity and increase team coordination by promoting common mental models across diverse crews.

A well-trained SOP becomes almost automatic, activating **procedural memory** – the brain's mechanism for skilled action (Squire, 2004). It's why a pilot can perform a recall item without conscious thought, or why a maintenance engineer can follow torque sequences accurately even while fatigued. Predictability turns uncertainty into rhythm, and rhythm is the mind's way of staying composed under strain (Hockey, 2013).

However, predictability must evolve with reality. EASA (2021) notes that static procedures can increase workload when they no longer fit modern operational contexts, such as mixed fleets or rapidly changing automation interfaces. The challenge is to preserve predictability without freezing flexibility.

Automaticity: A Double-Edged Sword

Automaticity has benefits – but also risks. As procedural tasks become ingrained, they often bypass conscious thought. That is efficient in stable situations but dangerous when adaptation is needed. A pilot can perform a checklist flow while chatting with Air Traffic Control (ATC), but if interrupted, that same automaticity can cause a skipped step.

Interruptions, stress, and time pressure can derail even well-rehearsed procedures. Worse still, the brain may fail to notice the lapse – a phenomenon known as cognitive tunnelling. The NASA Ames *Task Load and Error Recovery* study found that under acute workload, pilots often resume procedures from the wrong point after an interruption, with low detection rates unless verbal cross-checking is built in (Latorella, 1999).

A good example of this is **Singapore Airlines Flight SQ368 in 2016**. En route, the crew managed a right-engine oil/fuel anomaly and returned to Singapore. After touchdown, a post-landing fire developed around the right engine and wing; fire crews extinguished it within minutes. No emergency evacuation was performed; passengers disembarked via stairs after the situation was secured.

Airport fire crews responded promptly, though firefighting was briefly delayed as ground teams initially perceived the event as a routine checklist completion rather than a new in-flight emergency. The investigation found no procedural deviation, but highlighted how procedural focus can sometimes narrow situational awareness (TSIB, 2017)

It was a vivid example of automaticity doing its job perfectly – until reality changed faster than the checklist.

This interaction between automation, attention, and procedural habit is central to modern human-factors research. When operators overtrust routine or technology, manual skills and situational vigilance may erode. These themes will reappear in Chapter 10: *Manual Control, Automation, and Technology Hands-on Hands-off?*, where we explore how reliance on automation reshapes both procedural discipline and the human sense of control.

SOPs as a Cure for Decision Paralysis

Under extreme time pressure, SOPs protect against another common failure mode – **decision paralysis**. When multiple options compete for attention, the brain hesitates, looping between possibilities and delaying action (Klein, 1998; Orasanu and Martin, 1998). In aviation, that pause can be fatal.

A pre-agreed procedural rule – such as "rejected takeoff above 100 knots only for fire, engine failure, or unsafe configuration" – removes ambiguity and bypasses hesitation (O'Hare and Smitheram, 1995). The same principle applies when ATC controllers initiate emergency phraseology or when cabin crew open doors for evacuation. These are not robotic responses – they are neurological safety nets that let the brain act before panic overwhelms reason (FAA, 1994).

Research in naturalistic decision-making confirms this protective role. Klein's (1998) work on recognition-primed decision models and Orasanu and Martin's (1998) research on pilot judgment both show that pre-scripted procedural triggers dramatically reduce reaction time and cognitive load under stress. SOPs, in essence, give the brain permission to act decisively when time evaporates.

When Procedures Fit – or Fail – the Brain

Some of the most puzzling operational errors occur not from a lack of SOPs, but from their overuse. Pilots, controllers, and engineers sometimes continue applying standard steps because stepping outside them feels riskier than following them blindly. In psychological terms, this is **procedural fixation** – structure replacing sense-making (Dekker, 2017; Dismukes et al., 2007).

During takeoff from Tokyo Haneda on May 27, 2016, a **Korean Air Flight 2708** suffered a catastrophic failure of its No. 1 engine's low-pressure turbine disc, triggering an engine fire warning. The captain rejected the takeoff at high speed, brought the aircraft to a halt on the runway, and began the required memory items and checklists. Fire services were dispatched immediately.

While the crew followed the *Engine Fire/Severe Damage* and *Rejected Takeoff* procedures exactly as written, the unfolding situation demanded faster action. Flames and smoke were visible from the cabin, yet evacuation began only after outside confirmation from air-traffic control. Everyone survived, though several passengers were injured during the evacuation.

The Japan Transport Safety Board determined the cause to be a fatigue crack in the turbine disc and noted that, although procedures were correctly followed, aspects of coordination during evacuation and engine shutdown timing introduced additional risk (JTSB, 2018). The event became a vivid reminder that procedural fixation, continuing to execute the checklist when context has shifted, can delay critical actions. Sometimes, doing everything "by the book" can momentarily replace sense-making itself.

Neuroscience helps explain this trap. Under acute stress, the **prefrontal cortex**, responsible for flexibility and judgment, weakens, while the brain defaults to automatic routines (Arnsten, 2009). Structure becomes both anchor and cage. Engineers experience the same loop during troubleshooting, repeating steps that "should work" instead of questioning their assumptions (Dismukes and Berman, 2010).

Modern training now recognizes that procedural wisdom means knowing when to step outside the checklist. European Union Aviation Safety Agency's (EASA's) competency-based framework explicitly includes **adaptability and management of unexpected events** as assessable competencies (EASA, 2020). Flexibility is not the opposite of compliance; it is its mature form. Safety depends less on obedience than on awareness.

To support that awareness, **procedures must fit the brain**. The best ones are designed around how humans actually think, remember, and act under pressure. Research on cognitive ergonomics shows that short, clearly chunked steps improve recall and reduce error rates. Visual clarity matters too: consistent layout, logical grouping, and clear transitions between action and verification steps can reduce task time and errors (Degani and Wiener, 1993).

Strong SOPs also build in **sense-making pauses**: moments that invite crews to confirm whether the checklist still matches reality. These "cognitive stops" prevent blind compliance and restore situational awareness (Dismukes, 2010). It's the same logic as the **mental-health self-checklist** presented in Chapter 3 : *Mental Health and Well-being in Aviation: A Hidden Risk a Shared Responsibility*: a structured pause to verify not only system status, but human readiness.

Ultimately, procedures are not static rules but **adaptive frameworks**. They should support judgment, not suppress it; guide the brain, not govern it. When the SOP fits the brain, structure becomes liberation, not limitation.

AUTHORITY, CULTURE, AND GENERATIONAL NORMS

As we know already from previous chapters, not all cockpit behavior is driven by individual personality and attitudes. National and organizational culture play an equally powerful role – especially in how people perceive hierarchy, deference, and assertiveness.

National Cultures and Power Distance

In high power distance cultures – such as many parts of Asia or Latin America – junior pilots may hesitate to challenge captains or question ambiguous procedures. This hesitation can delay decision-making or lead to passive compliance, particularly under stress (Hofstede, 2001).

A telling example occurred during **Air India Express Flight IX1344** in 2020, arriving from Dubai, which attempted to land at Kozhikode, India, in heavy rain and variable wind. The aircraft was high and fast on approach. The first officer called for a go-around twice, yet the approach continued. The aircraft touched down long on the wet runway, overran, and broke apart on a slope beyond the threshold. Both pilots and 19 passengers lost their lives; more than 100 were injured.

The Aircraft Accident Investigation Bureau of India (AIBB India, 2021) found that the aircraft was serviceable and that the primary causes were continuation of an unstabilized approach and landing beyond the touchdown zone, compounded by weather and runway conditions. Among the contributing factors, the report noted aspects of crew resource management and assertiveness, highlighting the first officer's limited intervention and the captain's decision-making under stress.

While the report did not cite cultural factors directly, the event has been discussed in the context of **power-distance dynamics** – how hierarchy and deference can quietly influence procedural behavior. It shows how, even within standard operating frameworks, social norms and communication style can affect when and how challenges are voiced.

ICAO has long recognized the risks of such cultural gradients, and the introduction of CRM in the 1980s was intended, in part, to flatten these authority hierarchies. Yet culture remains a stubborn variable. The same SOP can be interpreted in completely different ways depending on local expectations about who leads, who follows, and how openly disagreement can be voiced (Figure 9.1).

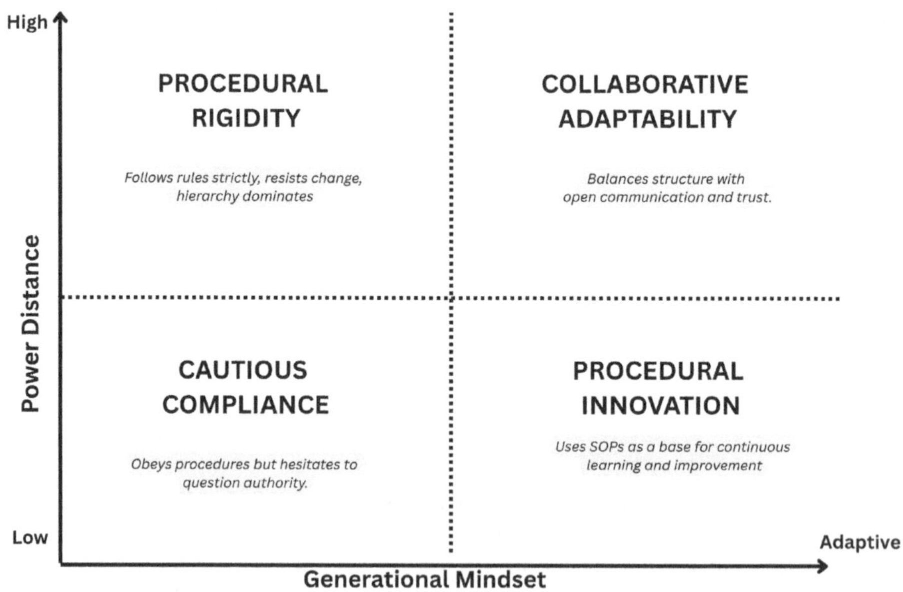

FIGURE 9.1 Cultural and generational influences on SOP interpretation.

Generational Shifts in Assertiveness and Trust

Inclusive SOP development helps achieve this balance. When pilots of different ranks, experience levels, and cultural backgrounds contribute to procedure design, compliance improves because ownership increases. Research shows that co-created procedures reduce resistance to change and strengthen safety culture by grounding rules in operational reality rather than purely regulatory logic (Maurino et al., 1995).

Instructors can reinforce this through discussion. Simple reflective questions, "How would someone new to this culture interpret this step?" or "Would this make sense to a newly qualified first officer?," build procedural wisdom that transcends both cultural and generational divides.

This interaction between power distance and generational mindset is illustrated in Figure 9.1, which shows how each combination produces a different procedural tendency.

DEVIATION, DRIFT, AND THE SLIPPERY SLOPE

Every flight sits somewhere between procedure and improvisation. Checklists guide actions, but judgment decides their use. Even experienced pilots sometimes bend a rule, skip a step, or adapt to what the SOP never anticipated. Such moments are part of aviation's reality – the negotiation between structure and sense.

Earlier chapters on Safety Culture and Decision-Making explored how judgment balances rules with real-world pressure. Here, the lens narrows to what happens when that balance tilts: when flexibility slides into drift and small shortcuts quietly redefine "normal."

The line between adaptation and erosion is thin. What begins as a pragmatic adjustment can, over time, reshape behavior and culture itself.

Not All Deviations Are Created Equal

Perfect adherence to SOPs is rare. Pilots work in shifting contexts – weather, traffic, fatigue, or conflicting ATC demands. A degree of adaptation is not only inevitable but often necessary.

The risk comes when these adjustments harden into habit. **Procedural drift** occurs when minor, harmless-seeming deviations are repeated until they become the local norm – what Dekker (2011) describes, at system level, as a **drift into failure**. It hides in patterns: a checklist item skipped because it "never matters," a callout delayed because "it's always fine." Distinguishing healthy flexibility from creeping complacency is the real skill.

When Deviation Helps Just Culture

Not every deviation signals weakness. Sometimes, it is necessary and reflects mature airmanship: structured flexibility grounded in awareness.

In 2018, **Sichuan Airlines Flight 3U8633** was cruising at 32,000 feet over China's Qinghai Plateau when the cockpit windshield shattered, causing explosive decompression. The first officer was pulled halfway out of the window, and most instruments and radios were degraded or temporarily unavailable. In the chaos, Captain Liu Chuanjian made a conscious decision to prioritize maintaining control of the aircraft before completing all checklist items.

He hand-flew the Airbus A319 in freezing wind and near-silence, descending and turning toward Chengdu. Only when the aircraft was stable and communications partially restored did the crew return to the checklist and coordinate with ATC. The Civil Aviation Administration of China (CAAC) later found that the crew's actions, including prioritizing aircraft control under extreme conditions, had been essential to the safe outcome. Rather than being criticized, they were publicly commended for professionalism and composure (CAAC, 2019).

This kind of attitude sits at the foundation of **Just Culture.** The key question is not whether a rule was followed word-for-word, but whether the choice was informed, transparent, and aligned with safety intent (Reason, 1997). A pilot who adapts for safety shows integrity; one who deviates for convenience erodes trust. Systems like Flight Data Monitoring (FDM) and Line Operations Safety Audits (LOSA) capture these real-world variations, so organizations can learn rather than blame.

A Just Culture does not demand perfect obedience, it seeks **intelligent compliance**: rules that guide behavior while judgment keeps them human.

Training to Recognize and Manage Drift

The best safeguard against drift is awareness, not control. No checklist can replace a mind trained to sense when adaptation becomes erosion (Dekker, 2011).

For instructors, teaching SOPs means exploring *why* they exist, *when* flexibility is justified, and *how* small deviations evolve into risk. Reflection and dialogue matter more than memorization (ICAO, 1998).

Modern competency frameworks such as **Competency-Based Training and Assessment (CBTA)** and **KSA100** already support this mindset. They assess not just actions but thinking: situational awareness, communication, and adaptability within boundaries (EASA, 2020). Embedding drift-awareness here turns safety into a living habit rather than a checklist item.

Organizations also play their part. Feedback loops that show crews how reports lead to change strengthen trust and openness (Dismukes and Berman, 2010). Ultimately, managing drift is about preserving intent. Procedures work when people understand their purpose and feel empowered to speak up the moment "normal" begins to slip.

WHEN SOPs DON'T EXIST – OR SHOULDN'T BE FOLLOWED BLINDLY

Even the most comprehensive manuals cannot imagine every scenario a crew might face. SOPs bring order to complexity, but real life has a habit of crossing their boundaries. The further aviation advances (with automation, new materials, and interconnected systems) the more likely it is that crews will face situations their checklists were never built to cover.

This section explores what happens when procedures fall short: when the script runs out, when the checklist doesn't match reality, or when following it to the letter could make things worse. It examines how procedural intelligence, flexibility, and training for ambiguity allow pilots and instructors to navigate these gray zones safely, turning uncertainty into awareness rather than panic.

Unscripted Situations in a Scripted World

Aviation prides itself on structure. Most in-flight events, from engine failures to pressurization loss, have checklists. But not everything can be scripted. Novel, ambiguous, or compound failures often force pilots into cognitive territory where procedures run out. In these moments, a well-trained procedural mindset becomes more valuable than any specific checklist.

Incidents involving volcanic ash ingestion, simultaneous system failures, or multiple conflicting alerts create "black swan" events: unexpected combinations of problems not directly covered by existing SOPs. the instinct to "find the right checklist" can become a cognitive trap, leading to fixation, delay, or even paralysis (Casner et al., 2013).

Sometimes, the SOP doesn't exist because the event has never happened before. Other times, a known checklist exists but doesn't match the actual failure or contradicts what the crew sees and feels. Blind compliance, in these cases, can escalate the risk rather than reduce it.

When Following SOPs Makes Things Worse

One of the most difficult human factors challenges is recognizing when a checklist – or part of one – no longer fits the moment. In high-stress situations, the instinct to follow procedure precisely can create danger instead of reducing it.

A memorable example came from **Japan Airlines Flight JL516** in 2024, an Airbus A350 that collided with a Japan Coast Guard aircraft upon landing at Tokyo Haneda. Within seconds, fire engulfed the fuselage. The evacuation checklist called for confirmation that engines were shut down, communication established with ATC, and slides armed before ordering passengers to exit. But the captain made a split-second decision to initiate evacuation rapidly as conditions deteriorated, prioritizing passenger egress under escalating fire and smoke conditions.

Every passenger and crew member escaped. The aircraft was destroyed, but all 379 people survived – a result investigators have preliminarily associated with timely evacuation and effective crew coordination rather than checklist fidelity. The Japan Transport Safety Board (JTSB, 2024) noted in its *preliminary findings* that the captain's actions reflected operational judgment under extreme time pressure, while the sequence remains under investigation, with rigid compliance potentially delaying evacuation.

This case illustrates an enduring truth in aviation psychology: checklists are designed for typical sequences, not catastrophic timing. When seconds separate safety from tragedy, pilots and cabin crew must prioritize *intent* – preserving life – over sequence. Judgment, teamwork, and shared understanding of the checklist's *purpose* become more critical than the order of its lines.

Instructors can use such examples to show that professionalism lies not in mechanical obedience but in knowing *why* each step exists, and when context demands adaptation (Kanki et al., 2019).

Procedural Intelligence: Thinking beyond the Page

Improvisation in aviation is not about abandoning procedures, it's about *thinking in their spirit.* **Procedural intelligence** or **intelligent procedural compliance** means understanding the goal of the checklist, not just the words on the page (Hollnagel, 2009).

It involves asking:

- *What is this checklist trying to accomplish?*
- *Which steps are essential, and which depend on assumptions that no longer hold true?*
- *What risks are added by following this step, and what risks are added by skipping it?*

These questions move aviation personnel from passive compliance to active decision-making – an approach endorsed in modern Crew Resource Management (CRM) frameworks (Salas et al., 2006).

Simulator training offers the perfect arena to build this mindset. Instructors can pause sessions to ask, "What made you follow that step?" or "What would you do if this procedure didn't work?" This transforms SOPs from static rules into dynamic tools for thinking.

DEVELOPING SOPs THAT WORK FOR HUMANS

SOPs are among aviation's greatest safety tools, but they are also psychological devices. They structure cognition when the mind is tired, distracted, or overloaded.

A well-designed SOP is not just a rulebook; it is a map of how humans think under stress. The best ones guide without overwhelming, define without confining, and leave just enough mental space for the most vital human skill of all – judgment.

Overconfidence and Familiarity: Dangerous Friends

Experience builds efficiency but can dull attention. Seasoned pilots often perform procedures automatically, assuming correctness through habit. Dismukes and Berman (2010) found that experienced crews under stable conditions skipped more checklist items than new crews, not from rebellion but routine streamlining. Efficiency, when unchecked, can turn into complacency.

SOPs should anticipate this drift toward automaticity. Regular cross-checks, standardized callouts, and intentional pauses before critical phases (such as "Before Takeoff" or "Before Landing") help restore focus. Procedures aren't meant to slow experts down but to keep them alert.

Designing for Cognitive Realism

Procedures are written by engineers but executed by humans coping with fatigue, noise, and time pressure. Degani and Wiener (1993) argued that checklists must reflect *how* work is done, not only how systems are configured. Human-centered design follows the same rule: align procedures with human logic, eye movement, and cockpit rhythm.

A checklist that works perfectly in an office may fail in turbulence or smoke. Testing SOPs under realistic simulator stress ensures they stay short, grouped, and clear about who does what. Well-designed checklists externalize memory, protecting cognition under pressure. Poorly written ones (too long, vague, or technical) can overload the brain when clarity matters most.

The Digital Flight Deck: When Procedures Go Paperless

Many airlines have moved their SOPs, manuals, and checklists to digital platforms such as **Electronic Flight Bags (EFBs)** and tablets. The benefits are clear: instant updates, reduced paper clutter, and seamless integration with performance calculators and weather tools. Yet, from a human factors perspective, digitalization introduces a new cognitive landscape (FAA, 2021; IATA, 2022; Joslin, 2013)

Digital SOPs are lighter to carry but heavier to navigate. Scrolling through a touchscreen in turbulence, interpreting complex sections, or waiting for an app to load adds subtle friction under pressure (Joslin, 2013). When a screen freezes or power fails, accessibility becomes a human reliability issue rather than a purely technical one.

Operational guidance for EFBs emphasizes usability, stability, and access paths that align with cockpit workflow – showing that effectiveness depends less on the technology itself and more on how well it supports mental flow (FAA, 2021; IATA, 2022). The most successful digital procedures mirror the clarity of well-designed paper checklists: short lines, clear titles, visible hierarchy, and intuitive touch paths

(Degani and Wiener, 1993). Poorly structured digital interfaces that bury steps or overload displays increase cognitive load and the risk of missed cues (Casner and Schooler, 2015; Hutchins, 1995; Norman, 2013).

Digitalization itself is neutral. Its value lies in how well it enhances the interface between human and system. Whether on paper or pixels, the guiding principle remains the same: *human logic first, format second* (Harris, 2011).

WHEN PROCEDURES AREN'T ENOUGH: THINKING BEYOND THE BOOK

Even the most carefully designed procedures cannot anticipate every operational reality. Aviation is dynamic, shaped by weather, infrastructure, human limits, and time pressure. Checklists and Standard Operating Procedures (SOPs) exist to reduce ambiguity and protect crews under stress, but they are built on assumptions: that systems behave as expected, that environmental margins are known, and that compliance reliably equates to safety. When those assumptions erode, strict procedural adherence may no longer be sufficient on its own.

Expert performance in such moments is not defined by abandoning procedures, but by understanding their intent. Skilled crews know when the task is to follow the book and when the priority must shift to preserving control, margin, or time. This is disciplined flexibility rather than disobedience – the ability to recognise when procedure supports safety and when it risks lagging behind a changing situation (FAA, 2016; Kanki et al., 2019).

When the Script Breaks

Standard operating procedures are designed to simplify complexity. They work best in familiar, repeatable conditions. However, every checklist reflects a model of the world that assumes predictability. When operational conditions diverge from that model, procedures may still be technically "correct" while becoming operationally fragile.

In such cases, crews face a cognitive challenge: recognizing that compliance alone may no longer guarantee safety. This requires stepping back from numerical thresholds or formal completion of steps and reassessing whether the underlying assumptions still hold.

In March 2015, **Air Canada Flight 624**, an Airbus A320-211, was conducting a night-time approach to Halifax Stanfield International Airport in winter conditions. During the final stages of the approach, the aircraft descended below the intended glide path and struck terrain short of the runway threshold. It continued through the localizer antenna array and onto the runway area, sustaining substantial damage. There were no fatalities, though 25 people were injured. The Transportation Safety Board of Canada (TSB, 2017) found no single technical failure, but identified environmental and perceptual factors that degraded the crew's ability to accurately assess the aircraft's vertical profile and energy state. Snow-covered terrain, reduced visual cues, and the night approach environment affected glide path perception. While the aircraft was serviceable, the event showed how environmental conditions can undermine assumptions embedded in approach guidance. The case illustrates how risk can emerge even when crews are managing a situation within an apparently familiar framework, and how safety can depend on recognizing when conditions reduce the reliability of those cues (TSB, 2017).

This event demonstrates a subtle but critical human-factors lesson. The procedures themselves did not fail, but the conditions reduced the reliability of the cues they depended on. As margins narrowed, the situation demanded not additional checklist discipline, but a reassessment of intent: whether continuing with a procedurally valid plan still preserved adequate safety buffers.

Thinking beyond the book does not mean improvisation without structure. It means recognising when the operational picture has shifted and being willing to pause, reassess, and adapt before procedural correctness becomes a constraint. In this sense, professional judgement is not an alternative to procedures, but their final safeguard. When crews remain alert to the limits of procedural models, they preserve safety not by abandoning the rules, but by understanding when the rules are no longer enough.

From Deviation to Wisdom

Deviation is often treated as an error, but in reality, it can be a proof of insight. Within a Just Culture, the key question is not whether someone deviated, but *why*. Was it reckless, or was it a rational adaptation to a situation the procedure couldn't predict?

When deviations are transparent and discussed openly, they become learning opportunities. They reveal gaps between design and operation, between system logic and human experience. As Reason (2000) reminded us, safety systems must "expect error, respect it, and design for it." That philosophy transforms deviations into data – lessons that refine SOPs and strengthen future performance.

Airmanship sits exactly where human adaptability meets procedural structure – echoing the integrated perspective of the SPINAL-CORD model introduced in Chapter 1. The procedure provides the frame, but wisdom lies in how humans interpret and apply it under pressure.

Some of the real-world events mentioned above – such as Cathay Pacific Flight 780 or Sichuan Airlines Flight 3U8633 – are not stories of rule-breaking, but of crews recognising when procedures alone were insufficient. In those cases, safety emerged from an understanding of intent: pilots adapted their actions while remaining anchored in system knowledge and shared reasoning.

Other events, such as Air Canada Flight 624, illustrate the complementary lesson from the opposite direction. Here, procedures were followed correctly, yet the assumptions embedded within them no longer aligned with operational reality. The outcome highlights that compliance, while essential, does not automatically guarantee safety when margins quietly erode.

Taken together, these cases show that professionalism in aviation is not defined by perfect adherence or deliberate deviation, but by the ability to recognize which mode is required. Procedures keep aviation predictable; judgment keeps it resilient. True expertise lies in knowing when the book is sufficient – and when it is not.

TAKEAWAYS

For Pilots and Cabin Crew

- Use SOPs as tools, not constraints. They are meant to guide and support (not replace) judgment, especially when stress or ambiguity rises.
- Speak up using the language of the SOP. Referring to a procedure helps challenge decisions without confrontation, particularly in high-authority settings.
- Respect the checklist even when it feels routine. Familiarity can dull vigilance; procedural discipline keeps awareness sharp.

Reflective Question: *When was the last time you used, or didn't use, a procedure to speak up, and how did it feel?*

For Instructors, Examiners, and Leaders

- Teach SOPs as cognitive tools, not compliance mandates. Explain why they exist, how they protect mental performance, and when thinking beyond them becomes necessary.
- Use simulator sessions to test boundaries. Include rare or ambiguous scenarios where students must balance procedural accuracy with sound judgment.
- Reinforce assertiveness through procedural language. Encourage pilots to challenge respectfully by framing feedback within the structure of SOPs.

Reflective Question: *How do you help cadets or new crew know when it's time to follow, and when it's time to think?*

For Operational Teams

- Design SOPs around human cognition. Use principles from psychology and ergonomics to shape logical task flow, memory cues, and interface layout.

- Build a culture that supports safe deviation. Recognize that flexibility is sometimes essential and create systems to discuss and learn from these moments openly.
- Avoid overloading SOPs with rare exceptions. Balance thoroughness with usability, and revise procedures regularly using feedback from the line.

Reflective Question: *Are your SOPs designed to serve the human – or is the human serving the SOP?*

REFERENCES

Aircraft Accident Investigation Bureau (AAIB India). (2021) *Final Accident Investigation Report on Air India Express Boeing 737-800, VT-AXH, Flight IX-1344, at Kozhikode, India on 7 August 2020.* New Delhi: Ministry of Civil Aviation, Government of India. Available at: https://aaib.gov.in/Reports/2020/accident/Final%20Report%20AXH.pdf

Air Line Pilots' Association Japan (ALPA Japan) (2014) *Cathay Pacific pilots receive IFALPA's Polaris Award* (Press release). Available at: https://alpajapan.org/cms_202306/wp-content/uploads/14PRL05-Cathay-Pacific-Pilots-Receive-IFALPAs-Polaris-Award.pdf

Arnsten, A. F. T. (2009) 'Stress signalling pathways that impair prefrontal cortex structure and function', *Nature Reviews Neuroscience*, 10(6), pp. 410–422. https://doi.org/10.1038/nrn2648.

Beringer, D. B., & Harris, Jr., H. C. (1999). Automation in general aviation: Two studies of pilot responses to autopilot malfunctions', *The International Journal of Aviation Psychology*, *9*(2), pp. 155–174. https://doi.org/10.1207/s15327108ijap0902_5

Bureau d'Enquêtes et d'Analyses (BEA). (2022) *Serious Incident to the Boeing 777 registered F-GSQJ operated by Air France on 05/04/2022 at Paris-Charles de Gaulle (93).* Final report. Le Bourget. Available at: https://bea.aero/fileadmin/user_upload/F-GSQJ_finalreport_EN.pdf

Casner, S. M., Geven, R. W. and Williams, K. T. (2013) 'The effectiveness of airline pilot training for abnormal events', *Human Factors*, 55(3), pp. 477–485. https://doi.org/10.1177/0018720812466893

Casner, S. M. and Schooler, J. W. (2015) 'Vigilance impossible: Diligence, distraction, and daydreaming all lead to failures in a practical monitoring task', *Consciousness and Cognition*, 35, pp. 33–41. https://doi.org/10.1016/j.concog.2015.04.019.

Civil Aviation Administration of China (CAAC) (2019) *Accident Investigation Report: Sichuan Airlines Flight 3U8633, Airbus A319-133, B-6419*, Chongqing-Lhasa, May 14, 2018. Beijing: CAAC. Available at: https://asn.flightsafety.org/reports/2018/20180514_A319_B-6419.pdf

Cowan, N. (2010) 'The magical mystery four: How is working memory capacity limited, and why?', *Current Directions in Psychological Science*, 19(1), pp. 51–57. https://doi.org/10.1177/0963721409359277

Degani, A. and Wiener, E. L. (1993) 'Cockpit checklists: Concepts, design, and use', *Human Factors*, 35(2), pp. 345–359. https://doi.org/10.1177/001872089303500209

Dekker, S. (2011) *Drift into Failure: From Hunting Broken Components to Understanding Complex Systems.* Farnham: Ashgate.

Dekker, S. (2017) *The Field Guide to Understanding 'Human Error'.* 4th ed. Boca Raton, FL: CRC Press.

Dismukes, R. K. (2010) *Monitoring, Teamwork, and Redundancy: Three Defensive Layers in Aviation Safety.* NASA Technical Memorandum 2010-216396. Moffett Field, CA: NASA Ames Research Center. Available at: https://ntrs.nasa.gov/citations/20100036673

Dismukes, R. K. and Berman, B. A. (2010) *Checklists and Monitoring in the Cockpit: Why Crucial Defenses Sometimes Fail.* NASA Technical Memorandum 2010-216396. Moffett Field, CA: NASA Ames Research Center. Available at: https://ntrs.nasa.gov/api/citations/20110011145/downloads/20110011145.pdf

Dismukes, R. K., Berman, B. A. and Loukopoulos, L. D. (2007) *The Limits of Expertise: Rethinking Pilot Error and the Causes of Airline Accidents.* Aldershot: Ashgate.

European Union Aviation Safety Agency (EASA) (2020) *Evidence-Based Training (EBT) – Guidance Material, Issue 1.* Cologne: EASA. Available at: https://www.easa.europa.eu/en/domains/aircrew-and-medical/evidence-based-training-ebt

European Union Aviation Safety Agency (EASA) (2021) *Annual Safety Review.* Cologne: EASA. Available at: https://www.easa.europa.eu/en/document-library/general-publications/annual-safety-review-2021

Federal Aviation Administration (FAA) (1994) *Advisory Circular 120-62: Takeoff Safety Training Aid.* Washington, DC: US Department of Transportation. Available at: https://www.faa.gov/regulations_policies/advisory_circulars/index.cfm/go/document.information/documentID/23202

Federal Aviation Administration (FAA) (2016) *Pilot's Handbook of Aeronautical Knowledge* (FAA-H-8083-25B). Washington, DC: FAA.

Federal Aviation Administration (FAA) (2021) *Advisory Circular 120-76E: Authorization for Use of Electronic Flight Bags (EFB).* Washington, DC: FAA.

Flin, R., O'Connor, P. and Crichton, M. (2008) *Safety at the Sharp End: A Guide to Non-Technical Skills.* Farnham: Ashgate.

Harris, D. (2011) *Human Performance on the Flight Deck.* Farnham: Ashgate.

Hockey, G. R. J. (2013) *The Psychology of Fatigue: Work, Effort and Control.* Cambridge: Cambridge University Press. https://doi.org/10.1017/CBO9781139015394

Hofstede, G. (2001) *Culture's Consequences: Comparing Values, Behaviors, Institutions, and Organizations across Nations.* 2nd ed. Thousand Oaks, CA: Sage.

Hollnagel, E. (2009) *The ETTO Principle: Efficiency-Thoroughness Trade-Off.* Farnham: Ashgate.

Hong Kong Civil Aviation Department (CAD) (2013) *Report on the Accident to Airbus A330-342 Aircraft B-HLL Operated by Cathay Pacific Airways Limited at Hong Kong International Airport on 13 April 2010.* Hong Kong: Air Accident Investigation Authority. Avialibale at: https://www.cad.gov.hk/reports/2%20Final%20Report%20-%20CX%20780%202013%2007%20web%20access%20compliant.pdf

Hutchins, E. L. (1995) *Cognition in the Wild.* Cambridge, MA: MIT Press.

International Air Transport Association (IATA) (2022) *EFB Implementation and Best Practices Guide.* 4th ed. Montreal: IATA.

International Civil Aviation Organization (ICAO) (1998) *Human Factors Training Manual* (Doc 9683, 1st ed.). Montreal: ICAO.

Japan Transport Safety Board (JTSB) (2018) *Aircraft Serious Incident Investigation Report: Boeing 777-3B5, HL7534, Korean Air Lines Co., Ltd., Tokyo International Airport, May 27, 2016.* Tokyo: Ministry of Land, Infrastructure, Transport and Tourism. Available at: https://www.mlit.go.jp/jtsb/eng-air_report/HL7534.pdf

Japan Transport Safety Board (JTSB) (2024) *Preliminary Report: Collision on Runway Involving Japan Airlines Airbus A350-900 JA13XJ and Japan Coast Guard DHC-8-300 at Tokyo International Airport on 2 January 2024.* Tokyo: Ministry of Land, Infrastructure, Transport and Tourism. https://jtsb.mlit.go.jp/eng-air_report/interim20241225-JA722A_JA13XJ.pdf

Joslin, R. E. (2013) 'Human factors hazards of iPads in general aviation cockpits', *Proceedings of the Human Factors and Ergonomics Society Annual Meeting*, 57(1), pp. 56–60. https://doi.org/10.1177/1541931213571015.

Kahneman, D. (2011) *Thinking, Fast and Slow*. New York: Farrar, Straus and Giroux.

Kanki, B. G., Helmreich, R. L. and Anca, J. (2019) *Crew Resource Management*. 3rd ed. San Diego, CA: Academic Press.

Klein, G. (1998) *Sources of Power: How People Make Decisions*. Cambridge, MA: MIT Press.

Latorella, K. A. (1999) Effects of Modality on Interrupted Flight Deck Performance: Implications for Data Link. *NASA Technical Memorandum 208894*. Moffett Field, CA: NASA Ames Research Center. Available at: https://ntrs.nasa.gov/api/citations/20000004291/downloads/20000004291.pdf

Maurino, D. E., Reason, J. T., Johnston, N. and Lee, R. B. (1995) *Beyond Aviation Human Factors: Safety in High Technology Systems*. Aldershot: Avebury Aviation.

Norman, D. A. (2013) *The Design of Everyday Things: Revised and Expanded Edition*. New York: Basic Books.

O'Hare, D. and Smitheram, T. (1995) '"Pressing on" into deteriorating conditions: An application of behavioral decision theory to pilot decision making', *The International Journal of Aviation Psychology*, 5(4), pp. 351–370. https://doi.org/10.1207/s15327108ijap0504_2

Orasanu, J. and Martin, L. (1998) 'Errors in aviation decision making: A factor in accidents and incidents', in D. J. Garland, J. A. Wise, and V. D. Hopkin (Eds.), *Handbook of Aviation Human Factors* (pp. 709–722). Mahwah, NJ: Lawrence Erlbaum Associates.

Reason, J. (1997) *Managing the Risks of Organizational Accidents*. Aldershot: Ashgate.

Reason, J. (2000) 'Human error: models and management', *BMJ*, 320(7237), pp. 768–770.

Reason, J. (2016) *The Human Contribution: Unsafe Acts, Accidents and Heroic Recoveries*. Farnham: Ashgate.

Salas, E., Wilson, K. A., Burke, C. S. and Wightman, D. C. (2006) 'Does crew resource management training work? An update, an extension, and some critical needs', *Human Factors*, 48(2), pp. 392–412. https://doi.org/10.1518/001872006777724444

Squire, L. R. (2004). 'Memory systems of the brain: A brief history and current perspective', *Neurobiology of Learning and Memory*, 82(3), pp. 171–177. https://doi.org/10.1016/j.nlm.2004.06.005

Transportation Safety Board of Canada (TSB) (2017) *Final Investigation Report: Airbus A320-211, Air Canada, flight AC624, approach and collision with terrain at Halifax Stanfield International Airport, March 2015*. Gatineau, QC: Transportation Safety Board of Canada. Available at: https://www.tsb.gc.ca/eng/rapports-reports/aviation/2015/a15h0002/a15h0002.html

Transport Safety Investigation Bureau (TSIB) (2017) *Final Report on the Engine Fire Involving Singapore Airlines Boeing 777-312ER, 9V-SWB, at Singapore Changi Airport on 27 June 2016*. Singapore: Ministry of Transport. Available at: https://asn.flightsafety.org/reports/2016/20160627_B77W_9V-SWB.pdf

Wickens, C. D. (2008) 'Multiple resources and mental workload', *Human Factors*, 50(3), pp. 449–455.

Wiegmann, D. A. and Shappell, S. A. (2017) *A Human Error Approach to Aviation Accident Analysis: The Human Factors Analysis and Classification System*. 2nd ed. Farnham: Ashgate.

10 Manual Control, Automation, and Technology

Hands-on or Hands-off?

CASE STUDY: ASIANA AIRLINES FLIGHT 214 (2013) – AUTOMATION, ILLUSION, AND MANUAL FLYING

It was a clear, quiet afternoon on July 6, 2013, when Asiana Airlines Flight 214 began its descent toward runway 28L at San Francisco International Airport. The Boeing 777-200ER, known for its advanced automation and reliability, carried 291 passengers and 16 crew from Seoul. Meteorological conditions met Visual Meteorological Conditions (VMC), and the flight was cleared for a visual approach. The ILS glideslope for Runway 28L was out of service, placing greater reliance on manual energy management and visual judgment. By all outward measures, it appeared to be a routine arrival. Yet within minutes, the flight would end in disaster.

In the cockpit sat three experienced pilots. The captain, over 10,000-hour veteran, supervised the pilot flying – another senior aviator completing his upgrade with nearly 9,700 hours. A relief first officer watched from the jump seat. Cleared for a visual approach, the crew expected to handle the descent manually with automation assisting. Expecting that the autothrottle would hold their target speed, the crew continued, unaware that pulling the thrust levers to idle had quietly placed the system in HOLD – a mode that no longer controlled airspeed (NTSB, 2014).

Airspeed began to decay. Still trusting the system, the pilots failed to intervene until the aircraft was dangerously slow. At 500 feet, the first officer realized the speed was far below target. Confusion and disbelief delayed their reaction. By 200 feet, with the runway rushing closer, they called for a go-around – but energy and altitude had already vanished.

The 777's tail struck the seawall, separating on impact and sending the aircraft spinning across the runway. Miraculously, the fuselage remained largely intact. Most passengers and crew escaped through smoke and debris, but three people were killed and 187 injured, many seriously (NTSB, 2014).

The investigation was sobering. Automation, designed to reduce error, had instead fostered overreliance and misunderstanding. Training and operational experience had not fully ensured that pilots consistently recognized subtle

DOI: 10.1201/9781003515418-10

mode changes or responded effectively when automation behaved unexpectedly. Manual handling, once instinctive, had become occasional. The National Transportation Safety Board (NTSB) concluded that mismanagement of automation and reduced hand-flying proficiency were central factors (NTSB, 2014).

The accident became a catalyst for revisiting training philosophy and redefining what true airmanship means in an automated age.

INTRODUCTION

In modern aviation, automation is no longer a new invention – it is now the backbone of flight operations. Pilots now command aircraft, which are capable of monitoring, navigating, and even landing with minimal human input. At the same time, as the Asiana 214 accident revealed, automation's quiet efficiency can mask its greatest risk. The challenge is not simply using automation, but knowing when and how to step back into full manual control, especially when the system might behave in an unpredictable way (NTSB, 2014; Wiener, 1989).

As aircraft systems developed, the pilot's role shifted from hands-on controller to system manager. Today's flight deck demands constant supervision of technology that quietly handles much of the workload. Paradoxically, while automation reduces human error, accidents linked to overreliance or misunderstanding continue to occur (FAA, 2013).

The central question of this chapter is deceptively simple: should pilots remain hands-on or hands-off? The answer, however, is layered. Manual flying skill fades without regular use, and modern line operations rarely require sustained hand-flying. The result is gradual skill erosion, reinforced by cognitive complacency that can weaken a pilot's ability to act decisively in high-stress situations (Casner and Schooler, 2014).

The arrival of Generation Z pilots, called by many "digital natives," fluent in technology, adds new complexity. They adapt quickly to intuitive digital systems but may need additional training to strengthen the perceptual, and kinesthetic skills essential for manual flight (Winter et al., 2019). Regulators and flight schools globally now face the same challenging task, which is blending modern digital training with the timeless craft of stick-and-rudder flying.

This chapter explores those tensions of the cognitive relationship between pilots and automation, offering research-based insights and practical strategies for pilots, instructors, and organizations to ensure aviation remains not only technologically advanced, but deeply human, adaptable, and ready for the challenges ahead.

THE EVOLUTION OF CONTROL: FROM STICK AND RUDDER TO FLY-BY-WIRE

The way pilots control aircraft has changed more in the last 15 years than in the previous hundred. Each step forward in technology brought smoother handling and greater safety, but also new mental demands. The shift from muscle to microprocessor replaced physical coordination with cognitive supervision. Understanding how automation reshaped flying means tracing how control itself evolved: from direct touch to digital trust.

A Brief History of Flight Control

Early aviators relied purely on muscle, intuition, and feel. A simple stick and rudder connected their movements directly to cables and controls. When autopilots first appeared in the 1930s, they offered relief from fatigue during long, monotonous flights (Jayawardana and Bandaranayake, 2021). Over the following decades, each generation of aircraft added another layer of assistance: radio navigation, flight directors, flight management systems, gradually moving the pilot from manipulator to monitor.

By the late 1970s, automation reached a turning point with the introduction of **fly-by-wire** technology. Mechanical linkages gave way to electrical signals, and computers began filtering every pilot input to keep the aircraft within safe parameters (Billings, 1997). The **Airbus A320** and later the **Boeing 777** symbolized this transition. Both used digital flight control systems capable of limiting unsafe attitudes, smoothing control responses, and integrating performance data in real time.

These changes made aviation safer than ever before. Yet they also shifted the nature of skill: from hands-on coordination to cognitive management. Instead of feeling aerodynamic feedback through control pressures, pilots began interpreting it through screens and logic displays (Wiener, 1989). The pilot's job was no longer to "fly the airplane" in a mechanical sense, but to manage a complex, self-regulating system.

What We Gained and Lost along the Way

Automation undeniably transformed flight safety. Controlled flight into terrain (CFIT) accidents dropped sharply with the introduction of ground-proximity warning systems and flight envelope protection. Long-haul operations became less physically demanding, reducing fatigue and extending performance.

But progress came with trade-offs. The same systems that improved reliability also reduced opportunities to practice core flying skills. Over time, manual handling became an exception rather than the norm. Many pilots today may hand-fly only a few minutes per leg. As a result, proficiency erodes, situational awareness drifts, and confidence in pure stick-and-rudder control fades (Endsley and Kiris, 1995; Sarter and Woods, 1995).

This tension between help and dependency is known as the **automation paradox**. The safer and more capable automation becomes, the less often pilots need to intervene. However, when something fails, those same pilots must act instantly and precisely, often after hours of passive monitoring (Evjemo and Johnsen, 2019). The Asiana 214 crash later illustrated this paradox very clearly: systems designed to simplify flying had instead impaired critical instincts when they were needed most (NTSB, 2014).

Automation has therefore shifted the pilot's greatest challenge from physical skill to **cognitive readiness**, so the ability to stay mentally connected even when the airplane is doing most of the work (Morrison and Fletcher, 2002).

Regulatory Shifts and Competency Expectations

Regulators have recognized this drift and responded. Both the International Civil Aviation Organization (ICAO) and the European Union Aviation Safety Agency (EASA) emphasize the importance of maintaining manual flying proficiency within Competency-Based Training and Assessment (CBTA) frameworks. ICAO's *Manual of Evidence-Based Training* (Doc 9995) highlights manual flight path management as a critical element that must be trained and assessed alongside automation management (ICAO, 2020).

EASA Safety Information Bulletin 2013-05 and Federal Aviation Administration (FAA) Safety Alerts for Operators (SAFOs) 13002 and 17007 similarly encourage airlines to allow periodic hand-flown segments during departure and approach phases, provided appropriate safety margins are maintained (EASA, 2025; FAA, 2013, 2017).

IATA's global survey of nearly 5,700 pilots found that about 69% reported their airline policy allows manual flying within specific limits, yet many still said they would like **more opportunities to hand-fly during line operations** (IATA, 2024).

The trend is clear: regulators and industry increasingly treat manual flying not as nostalgia, but as a critical safety layer. Maintaining manual competence isn't a luxury – it's insurance for when automation steps away.

Automation beyond the Flight Deck – Changing Minds across Aviation

The digital revolution didn't stop at the cockpit door. Every part of aviation, from control rooms to hangars and operations centers, has transformed. While the tools differ, the psychological challenge is the same: automation enhances performance but demands constant mental presence.

Air Traffic Control – Managing the Invisible

Controllers were among the first to face the paradox of automation. Early radar and conflict-alert systems eased workload but quietly changed attention itself. With modern tools and integrated flow management, controllers no longer "see" aircraft; they manage symbols and trajectories on screens (Wang et al., 2021). When these tools fail or alerts are suppressed, controllers must mentally reconstruct traffic – an exhausting task that exposes how deeply automation has rewired perception. Studies by Eurocontrol, NASA, and Eurocontrol/FAA collaborations show that long stretches of automated stability can erode detection skills and slow recovery when systems fail (Kirwan et al., 2010).

Maintenance Engineering – From Hands-On to Data-Driven

Maintenance engineers have followed a similar arc. Aircraft once spoke through sound and vibration; now they speak in data streams. Built-in test equipment and predictive analytics flag faults before touchdown. Reliability has improved, but the challenge is now cognitive: interpreting data critically rather than obeying it. In rare cases of sensor error or cascading faults, engineers must fall back on reasoning and inspection. Studies on maintenance human factors note that digital reliability can foster automation bias and skill fade if technicians lose tactile familiarity with aircraft systems (Reason and Hobbs, 2017).

Flight Dispatch and Operations – The Hidden Bridge

Dispatchers and operations controllers inhabit a world of near-total automation. Algorithms optimize routes, fuel, and alternates before human review. Nevertheless, every automated plan still meets weather, slots, and turbulence – and humans decide when the model no longer fits. Dispatchers act as a quiet safety layer, catching inconsistencies between digital systems. Research on operations monitoring highlights how situational awareness and cross-checking remain vital defenses against over-trust in flight-planning software (Salmon et al., 2009).

Across aviation, automation has done more than change procedures; it has changed how professionals think. Pilots, controllers, engineers, and dispatchers now manage complexity through abstraction: screens instead of gauges, algorithms instead of instincts. But the essence of airmanship remains: awareness, adaptability, and humility before uncertainty. Each role now practices a new kind of manual skill (*cognitive manual control*), the art of staying mentally hands-on in a system that rarely needs touching.

HUMAN-AUTOMATION INTERACTION: A COGNITIVE PERSPECTIVE

Automation has changed how humans think, not just how they act.

Pilots, air traffic controllers, engineers, and dispatchers now manage systems that do most of the physical work – but none of the responsibility. Each role faces the same invisible demand: to stay mentally hands-on in a world that rarely asks for physical touch.

Cognitive Demands and Vigilance Drift

Monitoring automation may look effortless, but cognitively it's among the hardest tasks in aviation. Research shows that prolonged system supervision depletes vigilance, known as ***vigilance decrement***, and reduces the likelihood of detecting subtle anomalies (Casner and Schooler, 2014; Warm et al., 2008). Many pilots describe cruise monitoring as paradoxically tiring: mentally draining yet physically inactive. The brain, denied meaningful engagement, struggles to stay alert.

Unlike manual flying, where sensory feedback sustains attention, automation relies on abstraction. A pilot reads numbers instead of feeling pitch or roll. Over time, this detachment erodes the intuitive link between mind and motion, producing cognitive underload. When underload meets fatigue, situational awareness fades quietly, often without the person realizing it (Endsley, 2016).

The same drift appears across aviation. Controllers watching steady radar tracks through a calm night describe how the picture "flattens," until one missing return jolts the mind back into focus. Maintenance engineers monitoring diagnostic feeds experience it too: hours of normal readings followed by one unexpected parameter that suddenly matters. Dispatchers, surrounded by stable data from flight-planning software, can also slip into passive acceptance until a fuel figure or wind estimate seems off. The human brain was designed for variation, not constancy, and automation's smoothness can lull any role into inattentiveness.

This cognitive drift has appeared repeatedly in high-profile events. The dynamics behind **Air France Flight 447** from 2009 (explained also in *Chapter 4: Stress and Resilience – Coping above the Clouds)*, where the autopilot disengaged and the crew misread the developing stall, illustrate how sustained monitoring, stress, and loss of direct sensory feedback can combine to degrade awareness (BEA, 2012). The same pattern appeared in the abovementioned **Asiana Airlines Flight 214** in 2013). Trust in the autothrottle led the crew to overlook a slow decay in airspeed during approach. What began as passive monitoring turned into confusion as mental models failed to update quickly enough (NTSB, 2014). In both cases, mental fatigue and expectation bias, not lack of skill, allowed automation to drift silently beyond human comprehension.

Mode Confusion and the Fragile Nature of Trust

If automation bias is blind trust, mode confusion is misplaced understanding. It occurs when humans believe the system is operating under one logic while it is, in fact, following another (Sarter and Woods, 1995). In modern aircraft, modes do not simply switch between "on" and "off," but transition across layered control laws, each governed by assumptions about sensor validity, redundancy, and priority. Pilots must reason not only about *what* the aircraft is doing, but *why* the system believes it should behave that way. When that reasoning diverges, trust becomes fragile (FAA, 2016).

Lufthansa Flight LH1829 in 2014 provides one of the clearest modern examples of exceptional airmanship in the presence of automation conflict. While cruising at altitude, the Airbus A321 experienced failures in two of its three angle-of-attack sensors. The flight-control computers, receiving contradictory inputs, interpreted the data as a stall condition and generated repeated nose-down pitch commands, despite the aircraft being aerodynamically stable (BFU, 2015).

Crucially, the crew were faced with a situation in which multiple sensors provided internally coherent but mutually inconsistent information. In such conditions, automated systems may behave correctly according to their logic while still producing unsafe outcomes, as the system is unable to determine which input reflects reality (Parasuraman and Riley, 1997). No amount of automated processing could resolve the disagreement because the problem was not computational but epistemic: the system could not know which sensor to trust (BFU, 2015).

Recognizing this limitation, the pilots demonstrated advanced systems reasoning rather than procedural reflex. They made a deliberate and informed decision to disable two flight-control computers, forcing the aircraft into Alternate Law. This removed the misleading automatic protections and restored full manual authority. The crew then stabilised the aircraft and continued safely.

This was not a loss of control followed by recovery. It was an intentional, reasoned disengagement from automation once its internal logic could no longer be trusted. Investigators later highlighted the crew's systems knowledge, composure, and collaborative reasoning as exemplary (BFU, 2015).

Across aviation (and other safety-critical domains), this pattern repeats. Whether it's a flight deck, radar console, or maintenance workstation, automation often reacts faster than human comprehension. The person supervising must decode intent – was that change a failure, a safeguard, or simply logic we forgot was active? When systems behave correctly according to incorrect inputs, trust fractures.

NASA research by Feigh and Pritchett (2019) found that pilots misunderstood one in four mode transitions during abnormal scenarios, particularly when visual cues were minimal. It's not a problem of competence but of cognition. Once the system begins thinking ahead, the operator becomes a step behind. This is the "out-of-the-loop" problem (Endsley and Kiris, 1995): the better the machine performs, the less we practice interpreting its behavior.

Training today increasingly targets trust assessments and teaching people to stay curious. That might mean verbalizing automation changes, cross-checking system outputs against expectation, or simply pausing to ask, "Does this make sense?."

Skill Decay and Cognitive Reversion: When the Machine Steps Back

When automation steps aside, the return to manual control is rarely graceful. It's not just the hands that must react – the whole cognitive system has to shift gears.

During normal operations, attention scatters across displays, communication, and planning. When automation disconnects, focus must snap back to raw data, muscle coordination, and spatial judgment. This reallocation takes precious seconds, creating the "attention-switching cost" (Wickens and Hollands, 2021).

Manual skill is invaluable. Psychomotor coordination, which is the subtle balance of pitch, roll, and thrust, relies on continuous sensory reinforcement. Without it, neural pathways weaken. Procedural knowledge may remain, but instinct fades. Many long-haul crews describe a quiet erosion of "feel" after months of monitoring rather than flying.

Other cases show what happens when automation not only steps back, but fights back. In **Ethiopian Airlines Flight 302** in 2019, The Maneuvering Characteristics Augmentation System (MCAS) repeatedly commanded nose-down trim, countering the crew's inputs. Each time the pilots trimmed manually, the system re-engaged. The battle between automation and human authority ended tragically when the aircraft impacted terrain (EAIB, 2022; see also NTSB, 2022).

Skill decay doesn't only erode physical control; it dulls confidence and timing. The same issue appears across roles, controllers reverting to manual sequencing after months of automation, engineers diagnosing without data support, dispatchers recalculating routes by hand. These acts may be rare, but they reveal whether the mind can still function without its digital scaffolding.

AUTOMATION SURPRISES AND THE STARTLE REFLEX

Automation rarely fails loudly. More often, it surprises quietly, like an unexpected pitch change, a sudden chime, a missing cue. These moments expose the fragile boundary between calm supervision and raw reaction. What happens next is not a checklist response but a human reflex centuries older than flight itself.

Automation-Triggered Startle and Pilot Response

The startle effect itself, its neurobiological underpinnings, contributing factors such as fatigue, and strategies for its management, is explored in greater depth in *Chapter 8: Attention, Fatigue, and Sleep – Balancing Act in the Sky*. Here, we focus specifically on how automation-related surprises can provoke a startle reaction and impair pilot performance.

When automation fails (or behaves unexpectedly), pilots are often faced with "automation surprises." These moments can trigger the **startle reflex:** a hardwired neurological and physiological response to unexpected stimuli (described in more details in Chapter 5). While startle can be adaptive in nature, helping us snap to attention, it also momentarily impairs fine motor control, decision-making, and working memory (Martin et al., 2012).

In aviation, this means that a pilot may freeze, misinterpret cues, or fail to take timely action in the seconds after an unexpected event. In the Asiana 214 accident, the crew's delayed recognition that the autothrottle had not re-engaged created a situation where the aircraft was far too slow on approach. The sudden realization triggered a partial go-around, but by then, the startle response had already disrupted their ability to problem-solve and coordinate effectively.

The effect isn't limited to pilots. Controllers facing a sudden radar dropout, maintenance engineers startled by an uncommanded activation, or dispatchers seeing a critical system crash – all experience the same physical jolt before cognition catches up. The challenge is not to eliminate startle, but to shorten its hold.

Research shows that the intensity of the startle response can be amplified by high workload, poor sleep, and low situational awareness: all factors often present in high-stakes flying (Vlaskamp et al., 2025). Even experienced crews can be caught off-guard by a momentary automation failure if they are not mentally prepared to intervene.

Delayed Action and Cognitive Overload

Surprise costs time. The mind first checks disbelief: *Did that really happen?* In those seconds, the situation evolves while cognition stalls. The Ethiopian Airlines Flight

302 crew faced exactly this overload, simultaneous warnings, stick-shaker activation, and conflicting data that demanded instant sorting. Even trained professionals can find such sensory storms impossible to prioritize.

Lesser-known events show the same delay. In earlier discussed case study from 2023: Austin runway incursion, FedEx Express flight 1432 on approach overflew Southwest 708 as the Southwest jet began its takeoff roll on the same runway; the local controller had incorrectly assumed the departure would be airborne before the arrival reached the runway in low visibility, prompting FedEx to go around at very low altitude (NTSB, 2024).

This hesitation is often called **confirmation delay**: the brain searching for proof that everything is still normal. In automated environments, that instinct can be dangerous, because the system keeps acting while the human is still reconciling the shock. It's a shared vulnerability across aviation roles; anyone who supervises technology long enough eventually meets the moment when it behaves unexpectedly (Landman et al., 2017).

Startle and delayed action are tightly linked. Training to mitigate these effects must go beyond traditional emergency drills and incorporate elements of uncertainty, degraded system cues, and the potential for misdirection.

Training for the Unexpected

Startle can't be erased, but it can be rehearsed. Modern human-factors training increasingly targets *surprise management* instead of rote failure response. Crews and controllers alike benefit from scenarios that blur the line between normal and abnormal, forcing awareness of emotional and cognitive reactions before technical ones.

Effective training programs teach a simple rhythm: breathe, verbalize, assess, act. This short pause regulates adrenaline and reopens working memory. Recent research shows that airline pilots using structured startle-management methods report clearer communication and faster recovery after unexpected automation events (Vlaskamp et al., 2025).

The foundation, however, remains understanding. Those who deeply grasp system logic, how protections work, what each alert really means, can recover faster. This is as true for a pilot reading mode logic as it is for a maintenance engineer interpreting an unexpected diagnostic code. Knowledge narrows panic's window.

Startle will always belong to aviation. What defines professionalism is not its absence, but the speed with which awareness returns.

Every decade brings faster machines and shorter reaction windows. The real challenge now isn't handling failure – it's managing **physiology**. The space between *what just happened* and *what I do next* is where safety lives.

Everyone in aviation eventually feels that instant when heart rate spikes and time seems to pause. The difference between chaos and control is simply who notices it first.

MANUAL REVERSION AND COGNITIVE READINESS

Automation has made aviation astonishingly safe, but safety still depends on what happens the moment the screen goes dark or unresponsive. Manual reversion isn't simply taking control again; it's reawakening the cognitive and emotional systems that automation had put at ease. Those moments are rare, but when they come, they demand everything at once: perception, coordination, and calm.

Automation-Out Scenarios: When Hands Must Take Over

Today's flight decks manage almost everything with precision: navigation, thrust, trim, even crosswind landings. Yet none of this is infallible. **Manual reversion**, which is the instant assumption of control when automation disconnects or behaves erratically, is no longer a backup event. It's a primary safety layer.

The expectation in modern operations is not that pilots fly less, but that they intervene faster and more precisely when automation crosses its limits. Reversion usually occurs at the most mentally loaded phases of flight: approach, climb, or a malfunction mid-air, when attention is already stretched. These are moments that test not only dexterity but situational composure.

In Asiana Airlines Flight 214, that transition came too late. The crew's deep trust in autothrottle logic delayed manual thrust input until airspeed was irretrievable. They knew the procedure, but not that the situation had crossed into danger. Their confusion wasn't a gap in skill – it was a gap in recognition (NTSB, 2014).

Similar moments unfold across aviation. Air traffic controllers revert to procedural separation after radar failure, engineers revert to manual fault isolation when digital diagnostics freeze, and dispatchers manually compute routes when flight-planning software misbehaves. In every case, success depends on how quickly the mind switches from expecting automation to noticing its absence. Manual reversion is therefore less about touch and more about mental elasticity.

Task Saturation and Mental Handoff

Today's flight decks manage almost everything with precision: navigation, thrust, trim, even crosswind landings. Yet none of this is infallible. Manual reversion, which is the instant assumption of control when automation disconnects or behaves erratically, is no longer a backup event but a primary safety layer (Casner and Schooler, 2014)

Neuroscience shows that humans don't truly multitask; we task-switch. Under stress, that switch can lag by seconds, which in aviation is a lifetime (Wickens and Hollands, 2021). A pilot who suddenly must fly, talk, and troubleshoot all at once faces the same cognitive bottleneck as a controller who must issue vectors manually or a maintenance engineer managing cascading alerts. The common risk is delay – the brain trying to "catch up" with reality.

Cognitive readiness is the antidote. It's built through anticipation – mentally rehearsing when and how automation might fail. Pilots brief "what if the autopilot drops out here," controllers imagine "what if the conflict alert freezes." Such mental simulations prime neural pathways, shortening reaction time when the unexpected happens (Landman et al., 2017; Morrison and Fletcher, 2002).

Checklists and Standard Operating Procedures (SOPs) provide structure, but the real advantage comes when response patterns become semi-automatic. Psychologists call this **automaticity under awareness** – smooth motor control supported by conscious monitoring that frees attention for decision-making (Endsley, 2016; Wickens and Hollands, 2021). It is not mechanical reaction but practiced calm.

As systems grow more complex, mental hand-offs become more demanding. Operators often face uncertainty not only about *what* failed but *why* – a form of cognitive ambiguity that taxes situational awareness and working memory (Feigh and Pritchett, 2019). The transition back to human control begins long before any alarm; it begins in a mindset that quietly assumes: one day, this system will surprise me.

STRATEGIES FOR BALANCING AUTOMATION AND MANUAL CONTROL

Automation is not the opposite of skill – it is its companion. The challenge is not to decide whether to fly manually or automatically, but to understand when and why to use each. The best aviators no longer think of control as binary. They navigate a spectrum of choices where automation becomes an extension of judgment rather than its replacement. True mastery lies in knowing when to let go and when to hold on.

THE "MANAGED USE" PHILOSOPHY

Modern human factors thinking encourages a philosophy known as **managed use** (Casner et al., 2013; Woods and Hollnagel, 2006). Rather than seeing automation as something to fear or blindly trust, managed use is about selecting the right level of automation for each phase of flight and situation. It is an active process, not a default mode.

This approach asks pilots to engage automation consciously and flexibly. During low-visibility approaches, automation may stabilize workload and increase precision. During dynamic weather, traffic patterns, or visual circuits, hand-flying often improves spatial awareness and re-engages attention. Automation should support the pilot's mental model, not define it.

The same mindset applies beyond the cockpit. Air traffic controllers decide when to rely on conflict-alert systems and when to intervene manually. Maintenance engineers balance diagnostic software with intuition built from experience. Dispatchers weigh automated flight plans against operational context. Each of these decisions represents managed use; the art of remaining mentally inside the system while leveraging its power.

Scenario-based training programs are increasingly embedding this principle. Pilots are asked to explain not only what automation level they selected, but why. This promotes a deeper kind of awareness – understanding the system's role rather than accepting it passively. Instructors encourage crews to treat automation settings as decisions, not habits. The result is a shift from compliance to conscious engagement, from automation as a shield to automation as a skill (EASA, 2021; ICAO, 2020).

At its best, managed use reflects modern airmanship: humans and machines sharing the same goal through deliberate cooperation, not default delegation.

Building an Automation Culture

Technology itself doesn't create complacency, culture does. Every organization shapes how its people relate to automation, often without realizing it. Studies of flight-deck culture show that when operational efficiency and fuel economy dominate conversation, pilots internalize that message and may view manual flight as deviation rather than discipline (Amalberti, 2001; Dekker, 2017). Over time, they default to the autopilot even when hand-flying would sustain sharper awareness.

A healthier culture values engagement. It recognizes that manual flying is not rebellion but resilience. When management, instructors, and safety departments reinforce curiosity about automation behavior, crews become confident partners rather than passive monitors (EASA, 2021; Endsley, 2016).

Effective automation culture is transparent, reflective, and non-punitive. It encourages the question "Why did it do that?" rather than the accusation "Who did that?" When unexpected automation events occur, the most constructive debriefs explore expectation gaps – where human assumptions and machine logic diverged (Dekker, 2012; Reason, 1997). These conversations build collective understanding instead of individual blame.

Operators can reinforce a balanced culture by:

- Encouraging hand-flying under stable, supervised conditions.
- Treating automation-related deviations as opportunities to learn.
- Providing clear, consistent guidance on when manual reversion is appropriate.

The Asiana 214 investigation highlighted how inconsistent automation training and unclear expectations about manual flight contributed to confusion (NTSB, 2014). The real issue wasn't procedural failure – it was more of a cultural one.

Culture does not shift through policy memos. It shifts through lived example. Every time a line captain hand-flies a stable approach, debriefs an automation event openly, or praises adaptability under uncertainty, the message is clear: automation is a tool, not a crutch. Engagement is professionalism, not defiance.

Instructor Modelling and Line Mentoring

The most powerful lessons about automation don't come from manuals, they come from people. Instructors and captains shape how pilots relate to automation through their own habits and language. Their actions set the tone for how an organization thinks and flies.

When experienced pilots explain their reasoning – why they engage or disengage automation, intervene or let it run, they reveal the invisible thinking behind skill. This transparency turns routine operations into mentorship moments. The next generation absorbs not only technique but mindset: curiosity, conscious decision-making, and respect for technology without dependence (Casner and Schooler, 2014).

Formal mentoring programs deepen this effect by creating space for dialogue about automation logic, human error traps, and recovery strategies. Observational learning, whether in simulators or jumpseat rides, lets junior pilots see judgment in action, not perfection but adaptability (EASA, 2021).

Effective instructors or managers don't just teach procedures; they teach flexibility. By introducing small anomalies or ambiguous alerts, they prompt students to interpret system behavior and verbalize expectations, building cognitive resilience. Instead of waiting for instructions, trainees learn to reason through confusion.

The best mentors model confidence without arrogance. They use automation strategically, stay calm when systems misbehave, and treat anomalies as puzzles, not threats. This attitude spreads through teams, shaping collective performance and situational awareness (Salmon et al., 2009). Automation is ultimately a mirror, it reflects the operator's habits. When leaders show curiosity and balance, those traits cascade. No technology upgrade influences safety as deeply as a human example that says: stay engaged, stay thinking, and never let the machine be the only one flying the airplane.

The relationship between technology, training, and human adaptability is not linear but dynamic. Each strengthens the others when held in balance. Overreliance on automation can dull manual awareness; rigid adherence to procedure can suppress initiative; but neglecting structure in favor of instinct creates chaos. True safety emerges when all three elements – system design, procedural discipline, and human adaptability, operate in equilibrium. This relationship is illustrated in Figure 10.1.

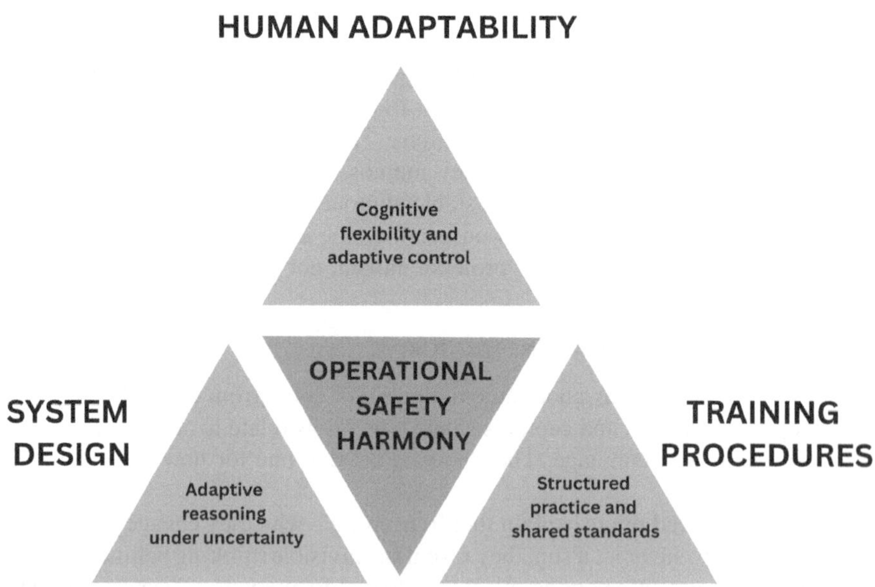

FIGURE 10.1 The balance of safety inputs.

TAKEAWAYS

For Pilots and Cabin Crew

- Treat automation as a tool that supports thinking, not a replacement for it.
- Use every safe opportunity to hand-fly and reconnect with the aircraft's natural feedback.
- Question automation states out loud: verbalizing keeps awareness active and prevents complacency.

Reflective Question: *When did you last deliberately disconnect automation to test your awareness and control?*

For Instructors, Examiners, and Leaders

- Model conscious use of automation – explain not only what you do, but why you do it.
- Design training that includes subtle surprises, degraded modes, and confusing cues to build adaptability.
- Debrief automation behavior as a shared problem-solving exercise, not an evaluation of error.

Reflective Question: *How often do your trainees see you balancing automation and manual control with clear intention?*

For Operational Teams

- Prioritize manual proficiency as a safety metric equal to technical accuracy.
- Encourage open reporting of automation surprises and moments of confusion without fear of criticism.
- Promote a culture where curiosity about system behavior is valued as professionalism, not doubt.

Reflective Question: *Does your organization reward those who question automation—or those who simply comply with it?*

REFERENCES

Amalberti, R. (2001) 'The paradoxes of almost totally safe transportation systems', *Safety Science*, 37(2–3), pp. 109–126. https://doi.org/10.1016/S0925-7535(00)00045-X

BEA (2012) *Final Report on the Accident on 1st June 2009 to the Airbus A330-203, Flight AF447.* Paris: Bureau d'Enquêtes et d'Analyses. https://bea.aero/fileadmin/uploads/tx_elyextendttnews/presentation.rapport.final.05juillet2012.en_04.pdf

Billings, C. E. (1997) *Aviation Automation: The Search for a Human-Centered Approach.* Mahwah, NJ: Lawrence Erlbaum Associates.

Bundesstelle für Flugunfalluntersuchung (BFU) (2015) *Serious Incident: Airbus A321-231, D-AIDP, Near Bilbao, Spain, on 05 November 2014.* Investigation Report No. 15-0025-AX. Braunschweig: German Federal Bureau of Aircraft Accident Investigation (BFU). Available at: https://www.bfu-web.de/EN/Publications/Investigation%20Report/2015/Report_15-0025-AX_Lufthansa_A321_Bilbao.pdf

Casner, S. M., Geven, R. W. and Williams, T. (2013) 'The effectiveness of airline pilot training for abnormal events', *Human Factors*, 55(3), pp. 477–485. https://doi.org/10.1177/0018720812466893

Casner, S. M. and Schooler, J. W. (2014) 'Thoughts in flight: Automation use and pilots' task-related and task-unrelated thought', *Human Factors*, 56(3), pp. 433–442. https://doi.org/10.1177/0018720813501550

Dekker, S. (2012) *Just Culture: Balancing Safety and Accountability.* 2nd ed. Farnham: Ashgate.

Dekker, S. (2017) *The Field Guide to Understanding "Human Error".* 3rd ed. Boca Raton, FL: CRC Press.

Endsley, M. R. (2016) 'From Here To Autonomy: Lessons Learned From Human-Automation Research', *Human Factors: The Journal of the Human Factors and Ergonomics Society*, 59(1), pp. 5–27. https://doi.org/10.1177/0018720816681350

Endsley, M. R. and Kiris, E. O. (1995) 'The out-of-the-loop performance problem and level of control in automation', *Human Factors*, 37(2), pp. 381–394. https://doi.org/10.1518/001872095779064555

Ethiopian Accident Investigation Bureau (EAIB). (2022) *Final Report: Accident to Boeing 737-8 (MAX), Registration ET-AVJ, Operated by Ethiopian Airlines, near Ejere, Ethiopia, on 10 March 2019. Addis Ababa:* Ministry of Transport and Logistics, Federal Democratic Republic of Ethiopia. Available at: https://avherald.com/files/ethiopian_b38m_et-avj_190310_final_report_20221226.pdf

European Union Aviation Safety Agency (EASA) (2021) *Guidance Material on Evidence-Based Training (EBT) – Issue 2.* Cologne: EASA.

European Union Aviation Safety Agency (EASA) (2025) *Safety Information Bulletin No. 2013-05R1: Manual Flight Training and Operations.* Cologne: EASA. Available at: https://ad.easa.europa.eu/ad/2013-05R1 (Revision R1, June 23, 2025).

Evjemo, T. E. and Johnsen, S. O. (2019) 'Lessons learned from increased automation in aviation: The paradox related to the high degree of safety and implications for future research', in *Proceedings of the 29th European Safety and Reliability Conference (ESREL 2019).* Singapore: Research Publishing Services. https://doi.org/10.3850/978-981-11-2724-3_0925-cd

Federal Aviation Administration (FAA) (2013) *Safety Alert for Operators (SAFO) 13002: Manual Flight Operations.* Washington, DC: FAA. Available at: https://www.faa.gov/sites/faa.gov/files/other_visit/aviation_industry/airline_operators/airline_safety/SAFO13002.pdf

Federal Aviation Administration (FAA) (2017) *Safety Alert for Operators (SAFO) 17007: Maintaining Proficiency in Manual Flight Operations.* Washington, DC: FAA. Available at: https://www.faa.gov/other_visit/aviation_industry/airline_operators/airline_safety/safo/all_safos/media/2017/SAFO17007.pdf

Feigh, K. M. and Pritchett, A. R. (2019) *Human-Automation Teaming: Lessons Learned and Future Directions.* NASA Technical Memorandum TM-2019-219937. Moffett Field, CA: NASA Ames Research Center. Available at: https://ntrs.nasa.gov/api/citations/20190001937/downloads/20190001937.pdf

ICAO (2020) *Doc 9995: Manual of Evidence-Based Training.* Montreal: International Civil Aviation Organization.

International Air Transport Association (IATA) (2024) *Aircraft Handling and Manual Flying Skills Report.* Montreal: IATA. Available at: https://www.iata.org/contentassets/d0e499e4b2824d4d867a8e07800b14bd/iata-report-aircraft-handling-manual-flying-skills.pdf

Jayawardana, R. and Bandaranayake, T. S. (2021) 'A review of the automated systems in civil aviation', *International Research Journal of Science and Technology*, 3(7), pp. 7–15. https://www.researchgate.net/publication/353700321_A_REVIEW_OF_THE_AUTOMATED_SYSTEMS_IN_CIVIL_AVIATION

Kirwan, B., Shorrock, S., Piccione, D. and Josefsson, B. (2010) *Human Performance in Air Traffic Management Safety: A White Paper. EUROCONTROL/FAA Action Plan 15 on Safety*. Brussels: European Organisation for the Safety of Air Navigation. 10.13140/RG.2.2.30236.97925.

Landman, A., Groen, E. L., van Paassen, M. M., Bronkhorst, A. W. and Mulder, M. (2017) 'Dealing with unexpected events on the flight deck: A conceptual model of startle and surprise', *Human Factors*, 59(8), pp. 1161–1172. https://doi.org/10.1177/0018720817723428

Martin, W., Murray, P. and Bates, P. (2012) *The effects of startle on pilots during critical events: A case study analysis*. 30th European Aviation Psychology Association Conference, Villasimius, Sardinia, September 24–28, 2012. Available at: https://www.researchgate.net/publication/275833443_The_Effects_of_Startle_on_Pilots_During_Critical_Events_A_Case_Study_Analysis

Morrison, J. E. and Fletcher, J. D. (2002) *Cognitive Readiness*. Alexandria, VA: Institute for Defense Analyses.

National Transportation Safety Board (NTSB) (2022) *U.S. Comments on Draft Aircraft Accident Investigation Report: Ethiopian Airlines Flight 302, Boeing 737-8 MAX, ET-AVJ, Ejere, Ethiopia, 10 March 2019*. Washington, DC: US Department of Transportation, National Transportation Safety Board. Available at: https://www.ntsb.gov/investigations/Documents/DCA19RA025_USComments.pdf

National Transportation Safety Board (NTSB) (2024) *Flight Path Management and Energy State Awareness: An Examination of Operational Safety Issues*. Aviation Investigation Report AIR-24/02. Washington, DC: US Department of Transportation, National Transportation Safety Board. Available at: https://www.ntsb.gov/investigations/AccidentReports/Reports/AIR2402.pdf

NTSB (2014) *Descent Below Visual Glidepath and Impact with Seawall, Asiana Airlines Flight 214, Boeing 777-200ER, HL7742, San Francisco, California, July 6, 2013*. AAR-14/01. Washington, DC: National Transportation Safety Board. https://www.ntsb.gov/investigations/AccidentReports/Reports/AAR1401.pdf

Parasuraman, R. and Riley, V. (1997) 'Humans and automation: Use, misuse, disuse, abuse', *Human Factors*, 39(2), pp. 230–253. https://doi.org/10.1518/001872097778543886

Reason, J. (1997) *Managing the Risks of Organizational Accidents*. Aldershot: Ashgate.

Reason, J. and Hobbs, A. (2017) *Managing Maintenance Error: A Practical Guide*. 2nd ed. Farnham: Ashgate.

Salmon, P. M., Stanton, N. A., Walker, G. H. and Jenkins, D. P. (2009) *Distributed Situation Awareness: Theory, Measurement and Application to Teamwork*. Farnham: Ashgate. https://doi.org/10.1201/9781315577654

Sarter, N. B. and Woods, D. D. (1995) 'How in the world did we ever get into that mode? Mode error and awareness in supervisory control', *Human Factors*, 37(1), pp. 5–19. https://doi.org/10.1518/001872095779049516

Vlaskamp, D., Landman, A., van Rooij, J. and Blundell, J. (2025) 'Recovery from startle and surprise: A survey of airline pilots' operational experience using a startle and surprise management method', *Aerospace Science and Technology*, 150, p. 109239. https://doi.org/10.1016/j.ergon.2025.103733

Wang, Y., Hu, R., Lin, S., Schultz, M. and Delahaye, D. (2021) 'The impact of automation on air traffic controller's behaviours', *Aerospace*, 8(9), p. 260. https://doi.org/10.3390/aerospace8090260

Warm, J. S., Parasuraman, R. and Matthews, G. (2008) 'Vigilance requires hard mental work and is stressful', *Human Factors*, 50(3), pp. 433–441. https://doi.org/10.1518/001872008X312152

Wickens, C. D. and Hollands, J. G. (2021) *Engineering Psychology and Human Performance*. 5th ed. New York: Routledge. https://doi.org/10.4324/9781003177616

Wiener, E. L. (1989) *Human Factors of Advanced Technology ("Glass Cockpit") Transport Aircraft*. NASA Contractor Report 177528. Moffett Field, CA: NASA Ames Research Center. Available at: https://ntrs.nasa.gov/citations/19890016609

Winter, S. R., Milner, M. N., Anania, E. C., Rice, S., Walters, N. W., Garcia, D. and Baugh, B. S. (2019) 'Examining the perception and effectiveness of a system awareness briefing during cruise flight', *Collegiate Aviation Review International*, 37(1), pp. 1–18. Retrieved from: Retrieved from https://ojs.library.okstate.edu/osu/index.php/CARI/article/view/7748/7192

Woods, D. D. and Hollnagel, E. (2006) *Joint Cognitive Systems: Patterns in Cognitive Systems Engineering*. Boca Raton, FL: CRC Press.

11 Future Ready
AI and Immersive Technologies in Aviation Human Factors

OPENING NOTE – WHY THIS CHAPTER HAS NO CASE STUDY

Every other chapter in this book begins with a moment that stopped the world: an engine failure, a runway overrun, or a call that came too late. We learn from what went wrong because aviation writes its lessons in black boxes. We also learn from what goes right – from the quiet saves, the skillful recoveries, and the teamwork that prevents small deviations from turning into huge disasters. As mentioned before, safety is not just the absence of failure but the presence of resilience (Hollnagel, 2014; Reason, 1997).

This chapter has no single key event because its story is still unfolding. Artificial intelligence (AI), immersive training, and predictive systems are already entering simulators, cockpits, and classrooms, but their safety record is still being written. We reached the times, where technology is moving faster than the human factors lessons that usually accompany it.

So rather than analyze an accident or incident, this chapter looks forward – to the human–machine partnerships now taking shape. It asks what kind of aviation culture we want to build and how we can design tools that strengthen, rather than replace, human capability. The future may be digital, but the guiding principle remains timeless: design for the human, not around them.

INTRODUCTION

Aviation has always evolved through innovation. From the first autopilot to fly-by-wire, each technological leap has rewritten what it means to be trained. Pilots learned to think in systems rather than instruments. Cabin crews became safety leaders, managing emergencies through digital checklists and crew communication tools. Air traffic controllers moved from paper strips to predictive radar, balancing trust in automation with real-time judgment. Maintenance engineers shifted from hardware to software, diagnosing faults from live data instead of worn parts.

Now a new wave of technologies is emerging – AI, virtual reality (VR) and augmented reality (AR), cognitive monitoring, or predictive analytics. They are faster, more adaptive, and more pervasive than any that came before. They do not just support decisions; they interpret, anticipate, and sometimes recommend. The challenge, as Parasuraman and Riley (1997) first warned, is not in building such systems but in keeping humans meaningfully in the loop.

DOI: 10.1201/9781003515418-11

This raises questions that go beyond technology and into psychology. How do we train pilots to challenge an algorithm that sounds certain? How do we help cabin crew preserve empathy when learning through headsets rather than human mentors? How do controllers stay vigilant when AI optimizes their workflow? And how do maintenance teams decide when to override a prediction that does not feel right?

These are not future hypotheticals. They are real situations unfolding in training centers and operations worldwide already (Cummings, 2014; EASA, 2023). This chapter explores how we can navigate them using the principles of human-centered design and modern human factors. The technology will continue to evolve, but safety will still depend on people who understand both: the system and themselves. In a sky shared with intelligent machines, human judgment remains our last line of reliability.

BEYOND THE COCKPIT – WHAT'S ALREADY HERE

When we talk about "future" technologies in aviation, it is easy to picture them as distant prototypes or some futuristic ideas. While in reality, many are already here. AI supports instructors in simulators. VR and AR reshape how crews train. Biometric monitoring appears in research and early trials.

Across aviation roles, this change is already visible. Pilots now interact with decision-support tools that scan parameters faster than they can. Cabin crew practice evacuations in 360-degree simulations. Controllers rely on predictive sequencing to prevent conflicts before they arise. Engineers consult models that predict faults before a wrench touches the aircraft. These technologies may still feel experimental, but they are quietly redefining how people learn, communicate, and make decisions.

How Artificial Intelligence Is Reshaping Training

European Union Aviation Safety Agency's (EASA's) AI Roadmap describes "Level 1" and "Level 2" systems that generate suggestions or highlight data, keeping humans firmly in charge (EASA, 2023). Similarly, the Federal Aviation Administration's (FAA's) *Roadmap for AI Safety Assurance* (2024) outlines an incremental "safety-continuum" strategy – starting with low-criticality use cases and expanding only as assurance frameworks mature – while stressing that AI must remain a tool under human supervision rather than an autonomous decision-maker.

The most common use in training is in simulator debriefing. Machine-learning algorithms flag moments of high workload or missed Standard Operating Procedures (SOPs), giving instructors quick access to key events (Yang et al., 2021). The aim is not to replace judgment but to focus it. The system identifies; the human interprets.

Physiological analytics are also emerging. By examining eye-movement patterns, blink rate, and electroencephalography (EEG), AI can reveal workload patterns and teamwork shifts in real time (Jiang et al., 2022). Used well, it gives instructors a clearer view of communication and stress. Used carelessly, it can misread style or culture as error. Similar caution applies in recruitment, where algorithmic analysis may promise objectivity while embedding hidden bias.

Across pilot, cabin, air traffic control (ATC), and maintenance training, the same rule holds: AI can help people see themselves more clearly, but only if they remain curious and critical. Once they stop questioning the tool, learning turns into compliance.

The New Partnership – Human Factors in AI-Supported Learning

The arrival of AI in training does more than add data; it changes how people trust information. Automation bias, the habit of deferring to a confident system, is well documented (Lee and See, 2004; Parasuraman and Riley, 1997). With AI, it becomes stronger because feedback appears intelligent and precise (Schaefer et al., 2016).

If an algorithm marks a "performance gap," instructors may accept it without context, overlooking fatigue, nerves, or workload. Likewise, a model that rewards "efficiency" might miss the teamwork and situational awareness that make operations safe. AI can measure what happened, but humans must still explain why (Cummings, 2014).

Avoiding over-trust means teaching instructors to question AI outputs, understand how algorithms draw conclusions, and validate them through behavior and judgment. Training organizations should build structured debriefs where humans and systems compare perspectives instead of one overruling the other.

Similar dynamics are emerging in other roles. Cabin crew trainers use AI to assess tone and clarity during safety briefings. ATC schools test systems that analyze phraseology accuracy in dense traffic. Maintenance courses use predictive tools to trace reasoning during fault-finding. In each case, the tool can help, until it starts deciding.

AI will make aviation training richer and more consistent, but its success depends on how humans frame it. The measure of progress is not how much the system can do, but how wisely we choose to use it.

Immersive Reality (VR/AR/XR) in Training

Immersive training is no longer a novelty. **VR, AR**, and **mixed reality (XR)** systems are now used across flight schools, airlines, and maintenance facilities to prepare crews in ways that traditional classrooms cannot (FAA, 2024b; Ross and Gilbey, 2023). The technology allows safe exposure to rare, high-stress events that would be too risky or expensive to simulate otherwise. Pilots practice cockpit procedures in full 3D, cabin crew rehearse evacuations with fire, smoke, and passengers moving realistically around them, and maintenance teams train on complex assemblies without grounding an aircraft.

Recent research, including a literature review on extended reality in aviation training by Klotz and Stendal (2023) and a 2024 FAA study (FAA, 2024a) on XR-based cabin-crew training, documents measurable benefits of VR in aviation learning environments. Crews reported feeling more immersed and better able to recall procedures, especially during emergency scenarios.

Some universities and aviation academies, including Embry-Riddle Aeronautical University, have implemented VR-based procedural and CRM training that allows instructors to pause, replay, and debrief interactive sessions using AI-assisted feedback (Ross and Gilbey, 2023; Thomas et al., 2022). The result is a more active and self-aware learning process.

The advantages, however, come with new human factors considerations. Extended headset use can cause fatigue and motion discomfort. Overexposure may even reduce transfer of learning if trainees begin to associate stress with virtual cues rather than real ones, as noted in early simulation research on stress exposure and transfer (Hart and Staveland, 1988; Ross et al., 2004). There is also the subtle risk of emotional detachment. A headset can simulate tension, but it cannot replicate empathy, social pressure, or the feeling of responsibility when real people are watching. To maintain realism, immersive sessions should always be followed by instructor-led reflection where emotions, teamwork, and decision-making are discussed openly. As with any tool in aviation, its value depends on how it is taught, supervised, and interpreted. The headset may provide the scene, but the lesson still comes from the conversation that follows

Alongside AI and immersive tools, another quiet shift is underway in aviation training and operations, which is **cognitive and physiological monitoring**. These systems measure what cannot be seen directly, like attention, workload, fatigue, and stress. In that way, what we used to learn from behavior can now be tracked in real time (Rice et al., 2025).

Eye-tracking, heart-rate variability, and EEG are among the most common indicators. In simulators, they reveal when focus fades or workload rises. Physiological indicators such as eye-movement patterns, heart-rate variability, and EEG can highlight rising workload or fatigue before clear performance effects appear (Caldwell et al., 2019). Trainers can then pause, debrief, and teach techniques to restore focus. Similar experiments are taking place in maintenance and ATC, where scanning patterns and fatigue profiles are being studied to fine-tune scheduling and workload distribution.

The benefits are clear: objective data can personalize training and help crews recognize their limits. But these insights also bring new ethical questions. If a headset can detect stress, who owns that data? Who decides what counts as "normal"? Privacy, transparency, and consent are essential (Cummings, 2014; Schaefer et al., 2016).

From a human factors perspective, physiological monitoring should support the operator, not monitor them. When introduced openly, with clear purpose and boundaries, crews tend to welcome it. When it feels intrusive or linked to evaluation, trust disappears.

Used wisely, cognitive monitoring can improve fatigue management, situational awareness, and personal resilience across all aviation roles. Used carelessly, it risks turning humans into data points. The technology's success will depend not on how much it can measure, but on how respectfully it is used.

RISKS, BIASES, AND HIDDEN HUMAN FACTORS

Innovation always brings promise, but also risk. AI, immersive simulation, and biometric monitoring may strengthen performance, yet they also create new blind spots. These risks are rarely mechanical; they are human. They arise from how people trust, interpret, and sometimes stop questioning technology. Human factors have always existed to explore these edges – the space between design and behavior, between intention and outcome.

Three risks stand out: automation bias, data misuse, and emotional detachment. Each has already appeared in early trials and will likely grow as technology becomes more deeply woven into training and operations.

Automation Bias in AI-Augmented Decision Systems

Automation bias is not new (we mention in many times in the book as well), but with AI, it gains a new layer of complexity. Traditionally, this bias describes a cognitive tendency to over-trust automated systems, even when evidence suggests they're wrong. In aviation, this might mean blindly accepting a route change offered by a flight management system, or misinterpreting a suggested SOP action without cross-checking context.

AI outputs often feel "intelligent" or "expert" – yet they remain non-transparent to many users. In safety-critical environments, this creates what Cummings (2014) describes as "automation over-reliance" and what trust researchers warn can become inappropriate reliance when feedback is ambiguous (Sellen and Horvitz, 2024).

Recent studies show that even highly trained professionals can overweight automated recommendations, especially when presented with confident or precise language (Lee and See, 2004; Schaefer et al., 2016). This is particularly concerning in training settings, where AI may be used to identify trainee errors or performance patterns. If instructors lean too heavily on algorithmic outputs, they may miss vital human context: nervousness, personality traits, or the subtle growth curve that only human observation can perceive.

Automation bias doesn't just affect individual decision-making – it affects how **teams** operate. In future human – AI collaborations, a "silent cockpit" might emerge, not from fear or hierarchy, but from quiet obedience to a digital assistant. This is why human factors must champion **explainability** and **challenge culture** in AI-augmented workflows (Endsley, 2017; Parasuraman and Riley, 1997).

Data Isn't always Objective

AI systems are trained on data – but data itself is not neutral. It reflects the values, assumptions, and blind spots of its creators. As Stinson (2021) notes, algorithms inherently reproduce the priorities and structures of those who build them, meaning "objectivity" in machine learning is more a design goal than a reality. When used in pilot selection, cognitive monitoring, or performance evaluation, algorithmic systems may encode unintended biases: against certain communication styles, accents, stress responses, or even learning patterns. For example, an eye-tracking system might mark frequent glances at an engine display as distraction, without recognizing that the pilot is **anticipating** a failure based on prior knowledge. A voice analysis algorithm might rate a quiet, calm tone as disengagement – when in fact it reflects strong emotional control. These systems can mislabel human variability as deficiency (Cañas et al., 2022).

Beyond the technical risk is an ethical one. If biometric and behavioral data is collected continuously: eye movement, EEG signals, heart rate, verbal tone – who owns it? Can it be used in disciplinary action? Could it unfairly influence upgrade decisions or crew pairing assignments?

From a human factors perspective, transparency and consent are non-negotiable (Cañas et al., 2022; Cummings, 2014). Otherwise, the cockpit becomes not just automated, but monitored in ways that **erode autonomy and psychological safety**.

Emotional Detachment or Emotional Support?

One of the more surprising frontiers of aviation training is the use of emotionally intelligent avatars and AI coaching agents. These systems aim to provide Crew Resource Management (CRM) feedback, assertiveness coaching, or post-scenario debriefs without a human instructor present. Recent work on emotional intelligence in artificial agents shows that deep multimodal models can approximate empathy through contextual adaptation and affective response (Annapareddy et al., 2025). Yet, as Cañas et al. (2022) and Cummings (2014) remind us, simulated emotion is not genuine understanding – machines can mirror tone, but they cannot share intent or care.

Studies in digital rapport-building show that while AI can simulate supportive language, it lacks the **dynamic empathy** that builds trust in high-pressure environments (Cañas et al., 2022). There is a growing concern that overuse of such systems could lead to **emotional detachment**, where learners become accustomed to sterile feedback, miss relational cues, or internalize a transactional view of safety culture (Endsley, 2017; Nizamani et al., 2024).

This is especially important for Gen Z learners, who are often tech-fluent but may struggle more with face-to-face conflict, critical feedback, or ambiguity. If their formative CRM training is delivered by avatars and checklists, rather than real human mentors, they may miss opportunities to develop **relational competence** – a skill that can't be coded, but only cultivated.

As Cummings (2014) notes, trust is not an interface problem. It's a relationship. Human factors must ensure that in our race toward efficiency, we do not automate away the very **connections** that make flying safe.

HUMAN–MACHINE COLLABORATION: THE NEW CRM?

When CRM was first introduced in aviation, it revolutionized how flight crews worked together – breaking down hierarchy, improving communication, and building shared decision-making under pressure. But CRM was fundamentally about humans working with humans. Now, a new type of team member is entering the picture – **digital, non-human, and increasingly intelligent**.

Whether it's a cockpit decision aid, a predictive maintenance algorithm, or a digital CRM assistant, AI is becoming a collaborative force in aviation. But as Sellen and Horvitz (2024) highlight, the design of AI co-pilots must prioritize transparency, trust, and shared situational awareness – principles long rooted in aviation but now essential across all human–machine partnerships.

In this section, we explore how the arrival of AI is changing the way pilots, cabin crew, air traffic controllers, and maintenance teams interact – not just with each other, but with the technology itself.

Human-in-the-Loop vs Human-on-the-Side

In aviation, the principle of "human-in-the-loop" has long been a cornerstone of safety. It ensures that a qualified person retains control, oversight, and authority over automated decisions. But as AI becomes more integrated, this loop is being stretched, and in some contexts, quietly reshaped.

For example, **cockpit AI assistants** are being trialed to support pilots with routing, checklist management, and emergency decision-making. Recent **adaptive/AI flight-control** research has informed demonstrator projects that aim to deliver more human-like, adaptive control responses – from NASA's neural-network-augmented Intelligent Flight Control System (IFCS) work to industry trials such as Airbus UpNext "DragonFly" (Airbus, 2023; Bosworth, 2007; Xargay et al., 2010). These systems are **not autonomous**, but their growing role raises the question: when AI makes suggestions, how does a human respond? Is it an aid – or a silent authority?

This issue isn't limited to the cockpit. In the cabin, AI-driven systems are being piloted to detect abnormal passenger behavior using sensors or facial expression tracking. Some prototypes alert crew to a possible unruly passenger before any verbal escalation occurs. But how do cabin crew interpret these alerts? Can they override them? And how might such systems influence crew judgement, empathy, or escalation strategies?

In air traffic control, AI-based tools are emerging to assist with sequencing, spacing, and conflict prediction. While these systems improve workload distribution, they also shift how controllers monitor aircraft. Some air traffic control officers report that AI-generated suggestions can become "too good," leading to complacency or reduced vigilance if not questioned actively (Lee and See, 2004). Others worry that decisions based on statistical confidence may lack operational context or local nuance.

Across the domains, the pattern is clear: AI is shifting roles from operator to **supervisor**. The human becomes the validator, the sense-checker, the final authority. But that authority only works if people are trained not just to use technology, but to question it.

As Endsley (2017) reminds us, maintaining shared mental models in human–automation teams is essential. These models must now extend across roles, and also across humans and machines.

Trust in the Process – The Missing Variable

Building calibrated trust between humans and intelligent systems is now one of the most critical frontiers in aviation psychology. Trust is not a simple matter of confidence in automation; it depends on transparency, competence, and feedback reliability. When any of these elements begin to fail, operators may either overtrust or **undertrust** a system – both of which degrade safety.

Figure 11.1 presents a conceptual model for understanding this balance. The *Human–Machine Trust Calibration Field* visualizes trust as a dynamic zone, where clarity of feedback, human expertise, and reliable system behavior converge to create optimal collaboration.

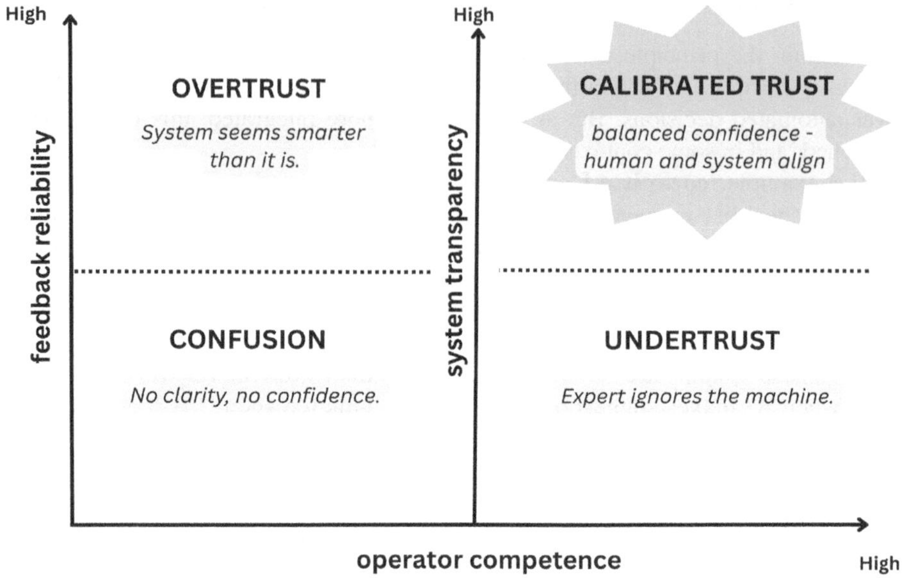

FIGURE 11.1 The Human–Machine Trust Calibration Field.

This model reframes trust as a *system property*, not just a personal attitude. When transparency is high but operator competence is low, overtrust can occur – a pilot accepting an AI recommendation without understanding its logic. Conversely, when competence is strong but feedback is unreliable, undertrust develops – a controller ignoring an accurate alert because previous ones were false.

Between these extremes lies the **Calibrated Trust Zone**: a psychological state where confidence matches actual system reliability. Achieving this balance demands more than good interface design; it requires AI literacy and cognitive awareness training. Instructors and safety leaders should specifically teach when and how to verify AI outputs, how to interpret uncertainty indicators, and how to question recommendations without undermining operational tempo.

The calibration of trust must therefore become a new dimension of CRM – the mental skill of working *with* intelligent systems rather than *for* them. It is the next step in aviation's evolution from human error prevention toward human–machine partnership.

From SOPs to SmartOps?

SOPs have long been the backbone of aviation safety. But in an AI-enabled world, there's a new idea emerging: **SmartOps**, which are adaptive procedures that respond to real-time context, workload, and performance data (Paksoy and Demir, 2024).

In the cockpit, AI systems could adjust checklist prompts depending on the phase of flight, aircraft configuration, or crew workload. If a pilot is overloaded during an emergency, the system might suppress non-critical items or reprioritize actions that are based on risk profiles (Endsley, 2017). But this requires deep trust in the system, and clear training on when and how to override it.

In the cabin, adaptive SOPs could theoretically support crew during disruptive events, suggesting calm-down scripts, rerouting decisions, or alternative service flows when delays or turbulence strike. But will these recommendations align with human intuition – or undermine it? And what happens if a junior crew member defers to the AI instead of following their training?

For air traffic controllers, smart procedures could reshape clearance flows. Imagine systems that automatically re-sequence traffic based on predicted trajectories, or propose dynamic reroutes during weather disruption. Controllers must remain alert, questioning whether AI suggestions fit their situational awareness, not just their radar screen.

When it comes to maintenance, SmartOps may generate repair schedules based on AI forecasts. While this increases efficiency, it also introduces **data dependence**. If maintenance teams begin to trust predictive diagnostics over their own inspection experience, mechanical insight could be lost, along with early indicators that fall outside of the model.

Whereas, in maintenance, SmartOps may generate repair schedules based on AI forecasts. While this increases efficiency, it also introduces data dependence. If maintenance teams begin to trust predictive diagnostics over their own inspection experience, mechanical insight could be lost, along with early indicators that fall outside of the model.

Recent reviews of automation and decision-support systems show similar effects across industries: efficiency improves, but humans gradually shift from active problem-solving to passive supervision, losing confidence in their own judgment (Eder et al., 2024). The human factors risk here is subtle: humans stop acting like experts and start acting like **executors**. Unless we actively teach crew, controllers, and engineers to critically engage with intelligent systems, SmartOps could become passive compliance systems rather than safety multipliers.

The next generation of CRM must therefore go beyond cockpit communication. It must evolve into cross-role, cross-technology resilience, where every team member is trained not just in what to do, but in how to think alongside AI.

DESIGNING WITH THE HUMAN IN MIND

Technology has always shaped aviation, but safety has never depended on technology alone. It has depended on how well that technology was designed for the people who use it. From the first yoke to the glass cockpit, from the checklist to CRM, every major advance in safety began not with a machine, but with an understanding of human limits.

Today, that foundation is under pressure. AI, immersive systems, and neurotechnology are entering aviation faster than the design principles meant to contain them. Some interfaces now seem built for data, not for people. Processes are optimized for algorithmic speed rather than cognitive clarity. When tools are designed around machines instead of minds, the human becomes the bottleneck – the one part of the system that no longer fits.

Human factors must therefore return to the design table not as a compliance step, but as a creative partner. Our role is not just to check whether a system is safe as such, but to ensure it *feels* usable, intuitive, and trustworthy to those who rely on it under stress, as only then it is truly safe.

Human-Centered Tech: Principles and Pitfalls

The central idea behind Human-Centered Design (HCD) is disarmingly simple: start with the user. Understand their environment, workload, and limitations, then build systems that complement their strengths and compensate for their weaknesses (Endsley, 2017). When that principle is followed, technology extends human ability. When it is ignored, it amplifies confusion.

In aviation, that means designing AI and automation that work *with* the operator, not *for* them. A well-designed interface reduces mental friction and supports attention rather than competing for it. Information should be clear, layered, and intuitive, while system feedback should be transparent enough that trust is earned, not assumed (Wickens et al., 2017).

The same principle extends beyond the cockpit. For cabin crew, digital service or safety tools must leave room for empathy and human judgment during real events. For controllers, alerts must align with scanning patterns and workload – too soon, and they distract; too late, and they overwhelm (Dekker, 2017). For maintenance teams, predictive systems must explain *why* a component is flagged, not only *that* it is, so that human trust and system logic evolve together (Lee and See, 2004).

When design skips the human, familiar problems return. A pilot is startled by an automation mode change. A controller misses a reroute because an AI system filtered it out. A mechanic ignores an alert after too many false positives. These are not random failures, they are artifacts of design processes that prioritized code over cognition.

Human factors specialists, especially those with operational experience, should be involved from the first sketch of any new system. Their presence ensures that safety is not an afterthought and that usability remains measurable, not assumed. As Parasuraman and Riley (1997) noted decades ago, human control is a design choice, not a given. The challenge nowadays is to keep making that choice.

Who Trains the AI?

AI does not arrive with wisdom. It learns from data, and that data must represent real human behavior, that is only complete with imperfections, emotions, and judgment calls. Without that, machines learn efficiency but not resilience (Cummings, 2014; Stinson, 2021).

In flight training, this means behavioral markers must be annotated carefully. AI must learn the difference between assertiveness and aggression, between hesitation and thoughtful caution. If data is labeled only by engineers, the system may reward task completion while missing signs of fatigue, tunnel vision, or poor teamwork (Rice et al., 2025).

The same risk appears in cabin and maintenance contexts. A model trained on service flows or defect logs without context may penalize adaptive thinking that kept operations safe. Machines see patterns, but only humans understand intent: a distinction central to ongoing work on ethical and transparent autonomous systems (Boddington, 2019).

So who trains the AI? The answer must include industry operators: pilots, instructors, controllers, engineers; also the specialist of human behavior and brain patterns – human factors specialists, physiologists, and psychologists who understand the cognitive and emotional layers of performance. Together, they define datasets, validate feedback loops, and challenge the model's assumptions to ensure the system captures the complexity of human decision-making under pressure (Amershi et al., 2019; EASA, 2023).

Human-AI collaboration must begin before deployment, at the dataset, at the algorithm, at the interface, to ensure that technology reflects not just computational logic but lived operational experience (Sellen and Horvitz, 2024).

The Cognitive Cost of Instant Thinking

The more capable our systems become, the less time the human brain spends on deliberate thought. AI now answers before we finish the question. Tasks that once required planning or reasoning are completed automatically. This instant efficiency feels empowering, yet it quietly reshapes how we think – from reflective to reactive, from understanding to executing. Recent studies show that frequent interaction with AI can alter attention regulation and emotional control, favoring speed over reflection (Nizamani et al., 2024).

Our brains were never built for this kind of tempo. Evolution equipped us for gradual analysis, not constant acceleration. The **prefrontal cortex** – the part responsible for decision-making, monitoring, and emotional regulation – depends on mental pauses. It needs friction, effort, and silence to make sense of complex problems. When technology removes those pauses, we lose the space where reasoning and judgment once lived.

This shift is subtle but profound. Over time, people can become conditioned to speed. They grow uncomfortable with uncertainty, intolerant of delay, and dependent on prompts to decide. The neural pathways that support sustained attention, working memory, and metacognition weaken from underuse. In aviation, where anticipation and self-checking are everything, that loss is not just cultural – it is operational.

AI can make us faster, but faster is not always better. Reflection takes time, and safety lives in that time. The checklist was never about speed; it was about mindfulness, a ritual to slow the brain just enough to think. When systems take over that role, crews may perform flawlessly until the moment something breaks, and they have forgotten how to slow down.

If technology is designed for instant response, humans must be trained for intentional thought. We need to teach patience as a safety skill, so the ability to pause, question, and resist the lure of the shortcut. Modern training must therefore cultivate cognitive friction, not eliminate it. In the cockpit, on the line, or in the tower, the most advanced system will still depend on an ancient truth: that humans think best when they are given the time to do so.

TRAINING THE NEXT GENERATION FOR A DIGITAL SKY

The next generation of aviation professionals is stepping into a world their instructors could hardly have imagined. Touchscreens, digital flight bags, voice assistants, and predictive algorithms are no longer novelties – they are part of everyday flying. Today's trainees are fluent in technology, but that fluency can be deceptive. Growing up with instant access to information does not mean knowing how to manage complexity, uncertainty, or failure (Kirschner and De Bruyckere, 2017).

For the first time in aviation history, technology is teaching people how to think as much as they are learning how to fly. AI tools summarize, suggest, and correct. They make decisions look effortless and remove much of the mental friction that once helped crews develop judgment. The result is a generation that is fast, connected, and adaptive – but sometimes less practiced in the slower, more deliberate thinking that safety demands.

Training in this environment must go beyond procedures. It must re-teach reflection, judgment, and patience – skills that technology quietly erodes.

Digital Natives ≠ Tech-Safe Operators

It is tempting to assume that those who grew up surrounded by digital tools are naturally equipped to use them safely. In many ways, they are. They learn new systems quickly, multitask with apparent ease, and navigate complex interfaces confidently. But digital fluency is not the same as digital maturity.

Research shows that so-called digital natives may be highly comfortable with technology, yet this comfort does not translate into better learning or decision-making. Kirschner and De Bruyckere (2017) found no evidence that growing up with digital tools enhances multitasking or cognitive control; in fact, frequent media switching can fragment attention and increase cognitive load. In other words, familiarity with technology is not the same as competence.

Recent studies suggest that constant interaction with AI tools further shapes cognition by rewarding instant feedback and reducing tolerance for ambiguity (Nizamani et al., 2024). Continuous digital cues train the brain to seek immediacy rather than understanding, which can erode the deeper reasoning and patience that aviation safety depends on.

Instructors now face a dual task: to develop this generation's comfort with technology while re-teaching them how to slow down. That means encouraging questioning, modeling how to challenge a system, and demonstrating calm reasoning when information runs out. The most advanced flight deck will still rely on the oldest human skill – judgement.

Mental Resilience and Cognitive Agility in Tech-Rich Environments

As automation grows, aviation roles are shifting from active control to active supervision. Pilots, controllers, and maintenance engineers now act more like systems managers – monitoring, verifying, and stepping in when needed. This requires a new balance of attention, adaptability, and emotional control.

Cognitive agility, so the ability to shift focus, update mental models, and recover from overload – is becoming as vital as technical skill. Studies in operational psychology show that expert performers maintain adaptability not by thinking faster, but by knowing when to slow down (Caldwell et al., 2019). They pause briefly, reassess, and realign before acting. Modern training should teach that pause deliberately through variable scenarios, surprise events, and reflective debriefs.

The same logic applies across aviation. Cabin crew now use AI-driven communication or safety tools that monitor passenger behavior. Controllers work with predictive systems that must still be cross-checked against instinct. Maintenance teams interpret automated fault reports that sometimes contradict what experience says. Across all these roles, people must remain the ones who make sense of it all. This shift mirrors what recent cross-domain reviews describe as a **supervisory drift**, where humans retain formal authority but gradually lose practical expertise through automation overtrust (Eder et al., 2024).

Resilience is what keeps that sense-making alive. As Endsley (2017) reminds us, situational awareness is not static – it is a constant process of updating one's understanding in motion. Training should therefore combine technical skill with mental awareness – teaching crews to recognize when technology supports them and when it quietly narrows their thinking.

The goal of training in a digital sky is not to outsmart machines, but to stay human beside them. The safest operators of the future will be those who understand both how technology thinks – and where it stops.

TAKEAWAYS

For Pilots and Cabin Crew

- Treat AI suggestions as part of CRM, not as authority. Verify digital inputs the same way you would a crew member's callout or decision.
- Keep your mental hand on the controls. Even in fully automated phases, continue running your own mental scan and anticipate what the system might overlook.
- Build deliberate thinking pauses into your routine. Before a descent, cabin service, or handover, take a brief moment to step back and reassess. Judgment often happens in the pause.

Reflective Question: *When technology makes something feel certain, what helps you stay curious enough to still question it?*

For Instructors, Examiners, and Leaders

- Model critical thinking around automation. Let trainees see how you question system outputs and explain your reasoning aloud.
- Teach AI literacy, not dependence. Help learners understand what algorithms do well – and where they tend to fail.

- Blend digital and human debriefing. Use simulator data or AI feedback as input, but connect it to behavior, context, and decision-making.

Reflective Question: *When discussing a trainee's error, do you focus more on what the data shows – or on why the decision was made?*

For Operational Teams

- Involve operators early in selecting or testing new technology. Including pilots, cabin crew, controllers, and engineers builds both trust and realism.
- Protect cognitive load in modern training. High fidelity is not always high learning; too many simultaneous feedback channels can reduce reflection.
- Ensure transparency in data use. If performance, biometrics, or behavior are monitored, communicate openly about what is measured, why, and how it supports learning.

Reflective Question: *Are your digital systems empowering people to think – or teaching them to wait for the machine to decide?*

REFERENCES

Airbus. (2023) *Airbus tests new technologies to enhance pilot assistance (UpNext "DragonFly").* Press release, January12, 2023, Available at: https://www.airbus.com/en/newsroom/press-releases/2023-01-airbus-tests-new-technologies-to-enhance-pilot-assistance

Amershi, S., Weld, D. S., Vorvoreanu, M., Fourney, A., Nushi, B., Collisson, P., Suh, J., Iqbal, S., Bennett, P. N., Inkpen, K. and Horvitz, E. (2019) *Guidelines for Human-AI Interaction. Proceedings of the 2019 CHI Conference on Human Factors in Computing Systems (CHI '19).* https://doi.org/10.1145/3290605.3300233

Annapareddy, V. N., Singireddy, J., Botlagunta, P. N., Lakarasu, P. and Burugulla, J. K. R. (2025) *Emotional Intelligence in Artificial Agents: Leveraging Deep Multimodal Big Data for Contextual Social Interaction and Adaptive Behavioral Modelling.* SSRN preprint, April 14. Available at: https://ssrn.com/abstract=5241039 or http://dx.doi.org/10.2139/ssrn.5241039

Boddington, P. (2019) 'Towards a code of ethics for artificial intelligence', *Delphi – Interdisciplinary Review of Emerging Technologies*, 2(2), pp. 105–106. https://doi.org/10.21552/delphi/2019/2/11

Bosworth, J. T. (2007) *Flight Test Results from the NF-15B Intelligent Flight Control System Project.* NASA/TP-2007. Available at: https://ntrs.nasa.gov/api/citations/20080001809/downloads/20080001809.pdf

Caldwell, J. A., Caldwell, J. L., Thompson, L. A. and Lieberman, H. R. (2019) 'Fatigue and its management in the workplace', *Neuroscience & Biobehavioral Reviews*, 96, pp. 272–289. https://pubmed.ncbi.nlm.nih.gov/30391406/

Cañas, J. J., Quesada, J. F., Antolí, A. and Fajardo, I. (2022) 'AI and ethics when human beings collaborate with AI systems: Accountability, responsibility, and co-supervision',

Frontiers in Psychology, 13, p. 836650. https://www.frontiersin.org/articles/10.3389/fpsyg.2022.836650/full

Cummings, M. L. (2014) 'Man versus machine or man + machine?', *IEEE Intelligent Systems*, 29(5), pp. 62–69. https://ieeexplore.ieee.org/document/6918133

Dekker, S. (2017) *The Safety Anarchist: Relying on Human Expertise and Innovation, Reducing Bureaucracy and Compliance*. Abingdon: Routledge/Taylor & Francis.

EASA (2023) *Artificial Intelligence Roadmap – A human-centric approach to AI in aviation* (Release 2). European Union Aviation Safety Agency. https://www.easa.europa.eu/en/document-library/general-publications/easa-artificial-intelligence-roadmap-20

Eder, K., Diederich, A. and Steinmann, S. (2024) 'Effects of automation and algorithmic decision support on human performance and trust: A systematic review', *Computers in Human Behavior*, 152, p. 108238. https://doi.org/10.1016/j.chb.2024.108238

Endsley, M. R. (2017) 'From here to autonomy: Lessons learned from human-automation research', *Human Factors*, 59(1), pp. 5–27. https://journals.sagepub.com/doi/10.1177/0018720816681350

Federal Aviation Administration (FAA) (2024a) *Extended Reality in Flight Attendant Initial Training* (DOT/FAA/AM-24/07). Washington, DC: U.S. Department of Transportation, Federal Aviation Administration, Office of Aerospace Medicine (Civil Aerospace Medical Institute). Available at: https://www.faa.gov/data_research/research/med_humanfacs/oamtechreports/media/202407.pdf

Federal Aviation Administration (FAA) (2024b) *Roadmap for Artificial Intelligence Safety Assurance*. Washington, DC: FAA. Available at: https://www.faa.gov/aircraft/air_cert/step/roadmap_for_AI_safety_assurance

Hart, S. G. and Staveland, L. E. (1988) 'Development of NASA-TLX (task load index): Results of empirical and theoretical research', in P. A. Hancock & N. Meshkati (Eds.), *Human Mental Workload* (pp. 139–183). Amsterdam: North-Holland. NASA archival copy: https://ia800504.us.archive.org/28/items/nasa_techdoc_20000004342/20000004342.pdf

Hollnagel, E. (2014) *Safety-I and Safety-II: The Past and Future of Safety Management*. Boca Raton, FL: CRC Press.

Jiang, G., Chen, H., Wang, C. and Xue, P. (2022) 'Mental workload artificial intelligence assessment of pilots' EEG based on multi-dimensional data fusion and LSTM with attention mechanism model', *International Journal of Pattern Recognition and Artificial Intelligence*, 36(11), p. 22590352. https://doi.org/10.1142/S0218001422590352

Kirschner, P. A. and De Bruyckere, P. (2017) 'The myths of the digital native and the multi-tasker', *Teaching and Teacher Education*, 67, pp. 135–142. https://www.sciencedirect.com/science/article/pii/S0742051X16306692

Klotz, A. and Stendal, K. (2023) 'How can extended reality (XR) enhance aviation personnel training: a literature review'. *NIKT*, no. 2, November 2023. Available at: https://www.ntnu.no/ojs/index.php/nikt/article/view/5672

Lee, J. D. and See, K. A. (2004) 'Trust in automation: Designing for appropriate reliance', *Human Factors*, 46(1), pp. 50–80. https://pubmed.ncbi.nlm.nih.gov/15151155/

Nizamani, M., Ramzan, F., Fatima, M. and Asif, M. (2024) 'Investigating how frequent interactions with AI technologies impact cognitive and emotional processes', *Bulletin of Business and Economics*, 13(3), pp. 316–325. Available at: https://bbejournal.com/BBE/article/view/1006

Paksoy, T. and Demir, S. (Eds.). (2024) *Smart and Sustainable Operations Management in the Aviation Industry: A Supply Chain 4.0 Perspective*. Boca Raton, FL: CRC Press.

Parasuraman, R. and Riley, V. (1997) 'Humans and automation: Use, misuse, disuse, abuse', *Human Factors*, 39(2), pp. 230–253. https://journals.sagepub.com/doi/10.1518/001872097778543886

Reason, J. (1997) *Managing the Risks of Organizational Accidents.* Abingdon: Ashgate (now Routledge). https://www.routledge.com/Managing-the-Risks-of-Organizational-Accidents/Reason/p/book/9781138252684

Rice, G. M., Linnville, S. and Snider, D. (2025) 'Methodologies using artificial intelligence to detect cognitive decrements in aviation environments', *Aerospace Medicine and Human Performance*, 96(4), pp. 327–338. https://doi.org/10.3357/AMHP.6555.2025

Ross, S. M., Szalma, J. L. and Hancock, P. A. (2004) 'The multi-dimensional structure of stress: Interactions among stressors and performance', *Aviation, Space, and Environmental Medicine*, 75(5), pp. C1–C6.

Schaefer, K. E., Chen, J. Y. C., Szalma, J. L. and Hancock, P. A. (2016) 'A meta-analysis of factors influencing the development of trust in automation', *Human Factors*, 58(3), pp. 377–411. https://journals.sagepub.com/doi/10.1177/0018720816634228

Sellen, A. and Horvitz, E. (2024) 'The rise of the AI co-pilot: Lessons for design from aviation and beyond', *Communications of the ACM*, 67(7), pp. 18–23. https://doi.org/10.1145/3637865

Stinson, C. (2021) *Algorithms Are Not Neutral: Bias in Collaborative Filtering. arXiv preprint*, arXiv:2105.01031. https://doi.org/10.48550/arXiv.2105.01031

Thomas, R. L., Carter, G. A., Nieves, N. A. and Barcza, T. D. (2022) Integration of Virtual Reality Procedural Training in a Flight Training Curriculum. https://commons.erau.edu/ntas/2022/presentation/11/

Wickens, C. D., Lee, J. D., Liu, Y. and Boyle, L. N. (2017) *Designing for People: An Introduction to Human Factors Engineering.* 3rd ed. Createspace/Independently Published. https://books.google.com/books/about/Designing_for_People.html?id=8gCTswEACAAJ

Xargay, E., Hovakimyan, N., Dobrokhodov, V., Kaminer, I., Gregory, I. M. and Cao, C. (2010) *L1 Adaptive Flight Control System: Flight Testing on a Subscale Civil Aircraft Model.* NASA/CR-2010-216251. Hampton, VA: NASA Langley Research Center. Available at: https://ntrs.nasa.gov/api/citations/20100017522/downloads/20100017522.pdf

Yang, S., Yu, K., Lammers, T. and Chen, F. (2021) 'Artificial intelligence in pilot training and education – towards a machine learning aided instructor assistant for flight simulators', in Stephanidis, C., Antona, M., Ntoa, S. (Eds.), *HCI International 2021 – Posters. HCII 2021. Communications in Computer and Information Science*, vol. 1420. Cham: Springer. https://doi.org/10.1007/978-3-030-78642-7_78

Index

E

F

G

H

I

J

K

L

M

N

O

W

Y

For Product Safety Concerns and Information please contact our EU representative GPSR@taylorandfrancis.com Taylor & Francis Verlag GmbH, Kaufingerstraße 24, 80331 München, Germany